Technocreep

and the Politics of
Things Not Seen

Technocreep

Contents

© 2025 DUKE UNIVERSITY PRESS. All rights reserved
Printed in the United States of America on acid-free paper ∞
Project Editor: Bird Williams
Designed by Courtney Leigh Richardson
Typeset in AdobeFnt24 and Minion Pro
by Westchester Publishing Services

Library of Congress Cataloging-in-Publication Data
Names: Atanasoski, Neda, editor. | Parvin, Nassim, [date] editor.
Title: Technocreep and the politics of things not seen / edited by Neda
Atanasoski and Nassim Parvin.
Description: Durham : Duke University Press, 2025. | Includes
bibliographical references and index.
Identifiers: LCCN 2024033419 (print)
LCCN 2024033420 (ebook)
ISBN 9781478031253 (paperback)
ISBN 9781478028031 (hardcover)
ISBN 9781478060239 (ebook)
Subjects: LCSH: Technology—Social aspects. | Technology—Political
aspects. | Technology—Economic aspects. | Technology and civilization. |
Art and technology. | Feminist theory.
Classification: LCC T14.5 . T4375 2025 (print)
DDC 303.48/3—dc23/eng/20250210
LC record available at https://lccn.loc.gov/2024033419
LC ebook record available at https://lccn.loc.gov/2024033420

Cover art: *Ineffable Freedom*, courtesy of Darya Fard.

and the Politics of Things Not Seen

EDITED BY NEDA ATANASOSKI NASSIM PARVIN

Duke University Press
Durham and London
2025

Prologue

This project started as a casual conversation at a coffee shop in Santa Cruz in 2018. Neither of us recall how this conversation began, but we remember that we were talking about everything from surveillance technologies deployed at the border and drones to smart fashion, smart trash cans, and the smart piggy bank that made its way into this volume. It was then that we first dreamed of the idea for this book, wondering if and how we could move away from the positivist and colonial histories that animate notions of what makes a technology "smart." We were looking for a more capacious sense and sensibility, a kind of knowing that we provisionally called feminist intelligence—an intelligence that is careful and caring, and at peace with unknowns and uncertainties. But as is with most dreams, this, too, was interrupted and truncated. We had to quickly return to all that demanded our attention. A whole year passed as we juggled new positions, tenure and promotions, unfinished projects, family and health. And then COVID-19.

Yet, the memory of the dream remained vivid. And its significance became even more apparent at the time of the pandemic, when we picked up the conversation again more than eight months into it. It was then and there that we took "the home" as a theme, along with technology, remembering the plight and bliss of digital technologies that shaped the home within and without, the borders and barriers technology made and remade, and the assumptions it made about insiders and outsiders. In our first Zoom meeting in the winter of 2021, we talked about home itself as a technology and revisited its technological extensions, cognizant of the inequalities they exacerbated and the possibilities they afforded. What does it mean to call a place home, anyway? And how can we expand and extend it as an idea and an ideal beyond its seeming commonsense borders and barriers?

It was home, then, that took us beyond its borders toward the discomforts and unknowns that it walled. And it was that move that made it clear to us that ours was a collective project, a many-voiced conversation that exceeded disciplinary homes and stubborn binaries. Our collaboration is deliberately opposed to disciplinary demands like "What is your percentage of contribution?" or "What section of the library will this book live in?" We were not writing to check some arbitrary box or fulfill an archaic and outdated academic demand. Rather, we wanted to write, we had to write, as a way of thinking, of making sense. There were pressing questions that we hoped to answer. Or even less ambitiously, perhaps, we were wondering how we could approach smart technologies in a capacious and generative way to invite a collective response commensurate with their complexity and urgency.

Two online panels and an open call later, the project took shape as the current volume. One of the key features of the volume is that it brings together scholars from a range of disciplinary backgrounds and employs a range of methodological approaches and genres of writing that span the artistic, speculative, ethnographic, and critical, among others. Our main driving ethos is that of community and conversation. We put forward the form of the book, the multiplicity of voices and media of expression it holds, as itself an intervention in the dominant modes of scholarship that tend to reward single voices asserting their authority. We hope that both this book's strengths and shortcomings serve as a beginning for new and different dreams and for polyphonic ways of dreaming our individual and collective futures.

Acknowledgments

As feminists, we know that dreams come to pass through collectivity. Thus, we have many people to thank and acknowledge. We thank all the contributing authors to this project, some of whom have been with us since nearly the 2021 symposium, for working with us through numerous drafts and revisions. We are grateful to Darya Fard for letting us use her stunning artwork, *Ineffable Freedom*, drawn from her exhibition *Myth Material*, as the cover of this book. The five visual artists whose work is included in the book have not only contributed their work but thought with us about technology, creep, and multisensorial ways of representing and rendering our research and arguments. Katherine Behar and the Artificial Ignorance group provided an engaging venue for us to share the concept of creep in relation to technology when the project was at an early stage. We are also grateful to Robert Rosenberger and students in the Fall 2023 Introduction to Science and Technology Studies (STS) course at Georgia Tech for their generous reading and feedback. Jordan Keesler's and Chris Fu's research assistance was invaluable as we brought the book to completion. We are grateful to Courtney Berger at Duke University Press for her early and insightful advice on producing a successful edited volume. Bird Williams was a joy to work with during the production stage. We would also like to thank our two anonymous reviewers who have made this a much stronger project with their thoughtful and incisive comments and suggestions.

Nassim

This book owes its existence to the incredible generosity and remarkable gift of Neda as a listener and collaborator. Her profound intellectual depth and curiosity have played a pivotal role in shaping this book. I am immensely grateful for her collaborative spirit and intellectual partnership and her everlasting imprint on my thinking.

This project spans multiple years, two institutions and many, many generous colleagues. I am forever grateful to my writing group friends Anne Pollock, Mary McDonald, and Jennifer Singh, who solidified my perspective on feminist ways of thinking and practice. My mentor and colleague Janet Murray has been a constant source of encouragement and inspiration. My friends Jennifer Anderson, Irina Aristarkhova, Kathrine Bennett, Matthieu Bloch, Carol Colatrella, Hayri Dortdivanlioglu, Laura Foster, Ian Hargraves, Soojin Jun, Sarah Kegley, Lisa Nathan, Vernelle A. A. Noel, Amit Prasat, Srirupa Prasat, Rebecca Rouse, Maryam Saeedifard, Leyla Salmassi, and Ren Zheng have given me the nourishment and love I've needed in various points of my life and scholarly career. Thanks to my former student turned friend Aditya Anupam for his generosity and the many inspiring conversations we've shared. I am deeply grateful to my colleagues and students at Georgia Tech, especially Charlie Bennett, Jay Bolter, Yanni Loukissas, Erica Taylor, Lewis A. Wheaton, and Richmond Wong. The warm welcome and support I received upon arriving at the school at the University of Washington gave me the energy I needed to finish this project in spite of the first few months of chaotic transition. Thank you, Cindy Aden, Marika Cifor, Anind Dey, Jin Ha Lee, Wes King, Amy Ko, Michelle Martin, Wanda Pratt, Daniela Rosner, Matthew Saxton, Maya Smith, Sasha Su-Ling Welland, Kimanthi Warren, and Shelly Wolf for your kindness and generosity. Heartfelt thanks to my life partner and friend, Azad. Thank you for so much love, for being my unwavering companion. To my beloved son, Kian, my wonderful sister, Ladan, and delightful nieces, Mahtab and Mahsa, thank you for filling my life with boundless joy! Thank you, dear Mom and Dad, Parvin and Jamal, for your enduring love and support that paved the path for me to pursue my dreams.

Neda

First and foremost, I would like to thank Nassim for her generosity as a thinker and collaborator. Our conversations provided much-needed space for creativity during a time of heavy administrative responsibilities. Nassim's love of impactful design and insistence that design, research, and theory cannot be separated will continue to inspire and guide my projects in years to come. I am grateful for my colleagues and friends at UC Santa Cruz, which is where this project started. Felicity Amaya Schaeffer has been an advocate of this book from the very beginning. Her friendship and insights into feminist technologies have nurtured this project over the years. Throughout the process of writing and editing this book, my *Critical Ethnic Studies* journal co-editor, Christine Hong, shaped my thinking about meaningful collaboration, and politically impactful scholarship. I enjoyed reading James Baldwin's *The Evidence of Things Not Seen* with Grace Hong and Rana Jaleel in fall 2019 tremendously, and it inspired the title of this book. My former students who are now my friends continue to energize me with their creativity and passion. Thank you, Erin McElroy, Francesca Romeo, Dana Ahern, Trung Nguyen, and Yulia Gilich—I always look forward to our conversations about both work and life! Dean Bonnie Thornton Dill, Dean Stephanie Shonekan, the College of Arts and Humanities, and the Artificial Intelligence Interdisciplinary Institute at Maryland (AIM) have generously provided support for this project since I moved to the University of Maryland, College Park, in 2021. My new colleagues in the Harriet Tubman Department of Women, Gender, and Sexuality Studies have been a joy to work with. Ruth Enid Zambrana, Neel Ahuja, Elsa Barkley Brown, Michelle Rowley, Alexis Lothian, Zenzele Isoke, Eva Hageman, Will Mosley, Sayan Bhattacharya, Shelbi Nahwilet Meissner, Sydney Lewis, and Jessica Lee Mathiason have modeled feminist collaboration that is at the heart of this project. I'm also grateful for friends in the larger UMD community who helped me settle into my new institutional home, especially Christina Hanhardt, David Sartorius, and GerShun Avilez. The multi-talented WGSS staff, Gwen Warman, Isata Munu, Jane Baitwa Kyomuhendo, and Elece Madison Smith, make it all possible. Hal Daumé III and Sheena Erete have been exceptional collaborators as we launched AIM while I was completing this book. I'm grateful to Marcus Gabriel and Christiane Schafer for the opportunity to be a Mercator Visiting Professor at the Center for Science and Thought at the University of Bonn, and for new colleagues there with whom I exchanged ideas by the beautiful Rhine River as this project was concluding. Friends in DC Caetlin Benson-Allott, Seth Perlow, and Sara Matthiesen make traveling to work something I look forward to. Longtime friends Rana Jaleel, Grace Hong, Kalindi Vora, Evren Savci, Heather Berg, Julietta Hua, Nina Martin, and Elizabeth Nagahara Boschee have continued to be there for me through many work and life transitions. Finally, without Neel Ahuja, my partner in everything, I could never have carved out the time to think and write. Thank you for always being present, for your caring and giving spirit, for your delicious cooking, for making me laugh, for all our adventures, and for your brilliance and warmth. You bring light and joy even to the most stressful times. Finally, this project could not have been completed without the love of my family: Neel Ahuja, Naya, Radoslav, Ljiljana, Vesna, Lulu, and Persia, who was with me at the start of the journey but not at the end. She is greatly missed.

NASSIM AND NEDA: What is a creepy technology?

CHATGPT: As an AI language model, I don't have preferences, feelings, or emotions.

However, *creepy technology* is often used to describe technologies that give people an uneasy feeling, invade their privacy, or seem too invasive. Examples might include facial recognition software that tracks people's movements without their knowledge or consent, or virtual assistants that listen in on private conversations even when not activated. Other examples include drones that can secretly follow people, wearable technology that tracks every movement a person makes, and social media platforms that collect personal data without permission.

The concept of creepy technology is subjective and can vary from person to person depending on their individual level of comfort with new forms of technology.

NEDA AND NASSIM: Are chatbots creepy?

CHATGPT: Some individuals may find them unsettling or eerie due to their ability to mimic human conversation, while others may not feel this way and find them helpful or interesting.

The perception of creepiness can depend on various factors, such as the design and behavior of the chatbot, the context in which it is used, and personal preferences. For example, if a chatbot uses realistic human avatars or displays behavior that is too lifelike, it might unsettle some people. On the other hand, if a chatbot is designed with a clear robotic or nonhuman appearance and behaves within appropriate boundaries, it may not be perceived as creepy.

It's worth noting that advancements in natural language processing and artificial intelligence have led to more sophisticated chatbots that can engage in more realistic and context-aware conversations. This can sometimes blur the line between human and machine interaction, which may contribute to varying perceptions of creepiness.

NASSIM AND NEDA: Are you creepy?

CHATGPT: As an AI language model, I don't have emotions or physical presence, so I don't possess the ability to be creepy. I'm here to provide information and assist with any questions or topics you'd like to discuss. How can I help you today?

Introduction: Technocreep and the Politics of Things Not Seen

What makes a technology creepy? What is at stake when a technology is deemed creepy? Could the answer to these questions be disruptive to how we imagine a good technology? Could creep itself unexpectedly serve as a starting point for imagining and enacting better relations with and through technology—ones that advance feminist and anticapitalist futures?

Think of two recent technologies: Qoobo (fig. I.1), the Japanese-designed headless furry robot with a heartbeat, produced and marketed to address loneliness and anxiety; and a small networked tooth sensor intended to seamlessly monitor food intake (fig. I.2). These examples conjure tacit concerns surfaced by creepy technologies: the increasing infiltration of technology into intimate realms, technologically induced isolation, the ubiquity and normalization of surveillance and the associated fears over the gradual loss of privacy, and the datafication and monetization of our lives, deaths, and afterlives. The tooth sensor, for example, typifies the biometric embeddedness of digital technologies in the human body and the idealization of the quantified self.[1] The headless robot, on the other hand, shores up an unease with accepting a companion that mimics life and appears animate but is not alive. Thus, unlike a stuffed animal, it may be viewed as further reinforcing solitude through a simulation of meaningful intimate relations.

When media outlets, designers, and engineers label a technology "creepy," its potential uses and harms are at once acknowledged and dismissed. As a descriptor of technology, *creepy* maintains the status quo around proper attachments and boundaries between the animate and inanimate, othering those with seemingly improper orientations to technological objects and platforms. And yet, we may wonder, would we still label it creepy if we realized that Qoobo is a source of comfort and companionship for a friend? The answer is likely not,

FIGURE I.I. Qoobo is a robotic toy with a heartbeat that reacts to human touch and sound, advertised as "a tailed cushion that heals your heart."

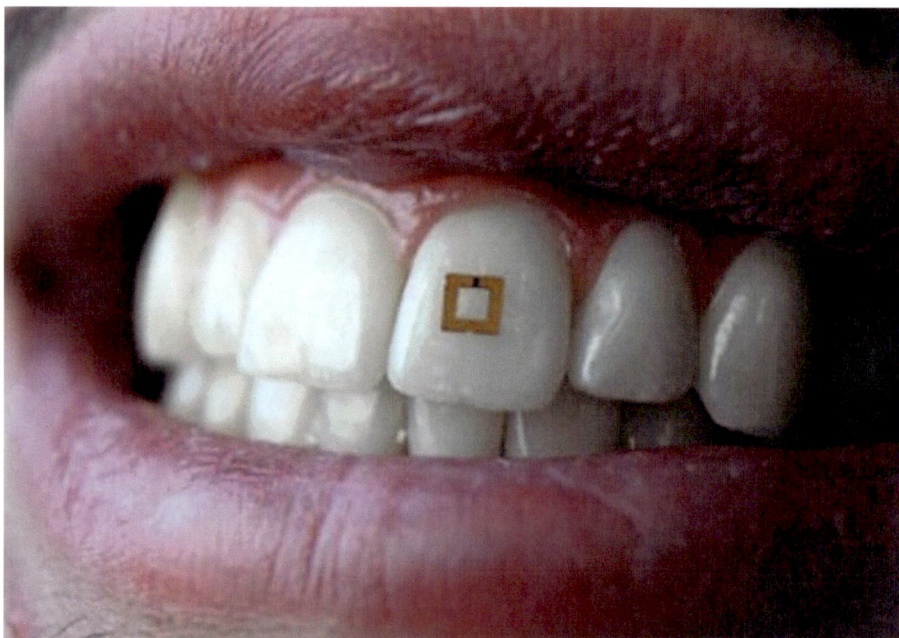

FIGURE I.2. The two-by-two-millimeter flexible sensor can bond to a tooth and monitor food and alcohol intake.

as doing so undermines the friend's judgment, emotional needs, or specific life situation. It would marginalize them and dismiss their affective bonds. The same may be said for the tooth sensor, which could indeed serve as an assistive technology, or address a medical need now or in the future.

New and emerging technologies, especially ones that infiltrate homes, bodies, and intimate spaces and relations, are often referred to as creepy in both public and scholarly discourse. Yet technological creep remains an undertheorized phenomenon, whether as a classification mechanism or a descriptive tool. This book sets out to render explicit what is otherwise obscured, assumed, or dismissed in characterizations of technology as creeping or creepy. As we write this book, beyond the most sensationalized emerging technologies and applications of artificial intelligence (AI), such as self-driving cars and deepfakes, the technologies most associated with creep and creepiness are those that are enmeshed in the fabric of everyday life. For the most part, they go unnoticed. Examples include but are not limited to smart speakers, CCTV cameras, body scanners (e.g., at airports), biometric devices (e.g., at the border), social media and the proliferation of fake news, and digital personal assistants that can compromise personal data.[2] As this list indicates, technological creep, or technocreep, is most commonly associated with surveillance and typified by its characterization as "a one way trip to the total surrender of privacy and the commoditization of intimacy."[3] Growing lists of creepy technologies thus act as a kind of shorthand warning about surveillance: "Products like Amazon Echo (powered by the Alexa personal assistant) and Google Home are popular household companions that respond to voice commands. But many experts are wary, citing the creepy behavior lurking just around the corner—like cases in which Amazon has already mishandled sensitive private recordings."[4]

When our daily habits are abstracted and commoditized as data, and our most mundane activities and social interactions in the digital realm become something that can be bought and sold, the fear of technological creep becomes about much more than surveillance—it becomes about the loss self-possession. In response, consumers' right to privacy is often posited as the best remedy to corporate overreach. Yet much is lost if we reduce the nexus of technology and creep to the loss of privacy. Instead, we argue, technological "creep" and "creepy" technologies mark the messiness of technologically mediated relations. Dwelling on those relations that cannot be described through binaries—such as privacy and surveillance, the public and the intimate, and harm and good—allows us to move beyond calls for a right to privacy as the only available politics of resistance that nonetheless accepts the technocapitalist present as a given. It instead allows us to move toward relations and politics that can disrupt racialized and gendered capitalist relations. In doing so, a feminist theorization of technocreep unsettles commonplace understandings of what makes a technology creepy and thus reorients us toward alternative possibilities.

To begin, we theorize four dimensions of creep that serve as the anchor for our analysis. First, *creep* describes a slow and unhurried movement that is often imperceptible. Geological creep is the gradual downward movement of soil along a slope. Creep becomes evident only after many decades, signaled by weathered, tilted gravestone markers on hills or slanted trees. Though imperceptible at first, this is a steady encroachment over time. Second, *creep* refers to the feeling that something potentially horrifying or repugnant, like a spider or worm, is moving over one's skin. Creep materializes with a sensation, a shiver, or a cringe. It is a sudden awareness of something that may not be seen yet is present—the creepy-crawlies. This aspect of creep names a mode of knowing that is about touch rather than sight. It is the realization of an unwelcome closeness or unthought intimacy. Third, *creep* describes the persistent growth associated with plants, especially creeping vines. In this mode, creep is a tenacious climb against odds. Vines "normally start by creeping along the floor. . . . Once they touch something, the physical contact triggers chemical changes that stimulate the climbing behavior and the plant begins to grow against the direction of gravity."[5] A tree, wall, or fence becomes the support structure that creeping plants need in order to grow. Though creeping vines can be seen as parasitic and colonizing, their creep could also be interpreted as mutual entanglement, endurance, growth, and life. Finally, *creep* is an expression of our intuition about something or someone being "off." It identifies a feeling that disturbing things loom ahead or lurk in the dark—things that may not yet be known or seen. The uncanny sense that all is not as it should be might gesture toward that which has the appearance of being harmless while being harmful and powerful, even menacing. Alternatively, that which may appear harmful, might yet prove harmless and friendly instead. Objects, animals, or people are characterized as creepy when they disturb commonplace assumptions and tacit or explicit presumptions of what is normal, such as when a line is unexpectedly blurred between human and machine, or animate and inanimate. What and who is framed as creepy is ambiguous, as are the situations under which some technologies, people, or ideas are labeled creepy.

This book draws on the rich interpretive, albeit ambiguous, capacity of *creep* to position it as a feminist method both for apprehending disturbances to normative relations that are valued under technologically augmented racial capitalism and for reorienting these relations toward justice-driven alternatives. We follow Jodi Melamed's definition of "racial capitalism," recognizing "that [all] capitalism is racial capitalism" and that "capital can only be capital when it is accumulating ... by producing and moving through relations of severe inequality among human groups."[6] Technocapitalism, which accounts for the ways in which the push to accelerate technological development and "innovation" works in the service of capitalism, builds on this scaffolding of racialized property relations. In opposition to the demands for acceleration in the service of efficiency and profit, we draw inspiration from creep's association with slowness and collaboration. We foreground ways of knowing and relating that are devalued and rendered invisible and incomprehensible by technological functions that prize speed, efficiency, and profit.

Technocreep and the Politics of Things Not Seen reclaims the messy contradictions of technologically mediated relations by dwelling on the temporal, spatial, felt, and normative dimensions of creep. Rather than categorize particular technologies as "good" or "bad," "useful" or "creepy," our book embraces the inherently relational nature of technologies as its core premise of theorizing. This allows us to think through the ethics and politics of technological relations in a manner that resists reducing all instances of technological creep to surveillance that can be solved through increased consumer privacy rights. We are critical of the conflation of technocreep with surveillance because the latter privileges sight. Instead, we argue that understanding the politics of technologically mediated relations demands a multifaceted and multimodal approach. Such an approach must take into account the kinds of experiences and modes of knowing that are inclusive of touch, sound, awareness, and intuition. It must account for relationships of trust, mutual understanding, and supportive interdependence.[7] For example, returning to Qoobo, we can consider how the headless cat raises questions about creep, intimacy, and our complex relations with technologies. We note that it is both headless and animate, yet it is not clear why. What may have happened to it? Because it is headless, the fact that it moves negates the normative but decidedly Western understanding of agency associated with the mind. Given that Qoobo does not have a head, how may we make sense of its movement, its uncanny imitation of life? Moreover, does a machine stepping into what is supposed to be a relation of intimacy and comfort conjure creep because it disrupts notions of how machines and humans are supposed to relate?

By dwelling on the nuances of what makes creep stick to some technological objects or platforms and not others, we might observe that creepy technologies have the potential to disrupt capitalist accounts of "good" human-machine relations that inherit racial and colonial demarcations of human and machine, in which the former is viewed as the commander and the latter is the subordinate. As ChatGPT put it in our opening interview, the commonplace understanding of what makes a technology creepy is its failure to "behave within appropriate boundaries." These boundaries uphold property and service relations to scaffold the operations of technologically augmented racial capitalism.[8] According to author 4 of the collectively authored piece "Making Kin with the Machines," "The Western view of both the human and non-human as exploitable resources is the result of . . . an 'epistemology of control' and is indelibly tied to colonization, capitalism, and slavery. Dakota philosopher Vine Deloria, Jr. writes about the enslavement of the non-human 'as if it were a machine.' . . . Slavery, the backbone of colonial capitalist power and the Western accumulation of wealth, is the end logic of an ontology which considers any non-human entity unworthy of relation."[9] By contrast, author 4 turns to Lakota epistemologies to ask how by "forming a relationship to AI, we form a relationship to the mines and the stones. Relations with AI are [also] relations with exploited resources." This approach to AI and technology sheds light not only on how technological relations are founded upon the exploitation of natural resources but also on how settler colonial and capitalist ontologies obscure these relations of production. Instead, "Indigenous ontologies ask us to take the world as the interconnected whole that it is, where the ontological status of non-humans is not inferior to that of humans."[10]

Throughout this book, we lean into the discomfort and unease of technological creep and creepy technology to think through human-technological interactions that have the potential to expose and unhinge hierarchical and exploitative relations. We embrace the temporal and spatial multimodality of creep to challenge what and how we know. The slowness of creep works against capitalist drives to continually innovate technological development in a push for apparent newness. Thus, as a feminist method of slowing down, creep moves our analysis away from constantly following the newest technologies and allows us instead to dwell on the histories that undergird our present moment. In doing so, this edited collection takes up creep as an analytical and creative mode of engaging technologies' entanglements with both the intimate and local alongside the all-encompassing and global.

THE TIMESPACE OF TECHNOCREEP

"The 11 Creepiest Technologies That Exist Today"[11]

"10 Scary Modern Technologies"[12]

"9 Terrifying Technologies That Will Shape Your Future"[13]

"GPT-4 Is Exciting and Scary"[14]

This list of titles represents just a few of the myriad articles warning of an exceptional and novel order of things ushered in by creepy technology. Technologies that are viewed as creepy in the mainstream press occupy a curious sense of time—always already here but also not yet here. The advent of new, ever more creepy technologies is perpetually announced. Their seemingly inevitable existence forecloses any possibility of resistance or redirection. We cannot possibly do anything about them but accept their inescapable arrival, as it is already too late anyway.

Refusing resignation as an option, in this book we ask instead, What does it take to expand on creep as an analytical starting point for apprehending complex and contradictory technological entanglements with human and more-than-human worlds? What is lost, forgotten, or neglected in the insistence that technological creep is unprecedented, new, and uniquely threatening? Could rendering technology and technological relations creepy obscure violence and neglect (with life-or-death consequences) or alternatively dismiss or negate lives and livelihoods at the margins? We find the approach of naming, classifying, and charting creepy technologies as tools for surveillance and spying, on one hand, or for taking away our "humanity," on the other, inadequate and even harmful. These rhetorical gestures deflect from the nuances that so-called creepy technologies surface. For us, the conceptual ambiguity of technocreep renders it an apt feminist method for foregrounding the unseen, felt, and otherwise underappreciated and untheorized dimensions of technology. Creep's association with plants, for instance, is partly an invitation to consider how digital technologies affect more-than-human worlds and human relations. It thus transcends the presumed human-machine divide in which creepy technologies threaten our humanity—as if the fully human subject wasn't always a construct produced over and against the other, the monstrous, the less than human, and the inhuman.

As an analytic, technocreep disrupts how we assess "intelligence" by foregrounding intuition and multimodal forms of knowing associated with creep. The multimodality of perception inherent in what creep names avoids the pitfalls of reducing technological relations to axes of seeing and being seen (the visual) and instead highlights other sensorial dimensions that are integral to our ways of knowing. For example, as we argue in one of this book's interludes, a feminist approach to the design and application of smart forest technologies (sensors deployed in the woods) refashions how we understand our place and relate to the more-than-human.[15] Rather than studying creepy technologies, we move toward using creep to open up feminist ways of knowing, relating, seeing, and sensing with and against technology.

Refusing to join in the catastrophizing and disempowering rhetoric around creepy technologies as a new and exceptional phenomenon, our title, *Technocreep and the Politics of Things Not Seen*, references James Baldwin's 1985 essay "The Evidence of Things Not Seen" about the spate of murders suffered by

Black children in 1980s Atlanta.[16] As a recent *New Yorker* magazine retrospective of the essay states, Baldwin "marshals the injustice of one set of cases not only or even chiefly to resolve them but in order to make an argument about justice itself. 'The Evidence of Things Not Seen' is less a book about the deaths of black children by helping us see them and their lives—inclusive of the violence and neglect that too often afflict them and about the ways that, in today's parlance, they do and do not matter."[17]

Writing in the midst of extensive and sensationalizing coverage of the child murders, Baldwin's focus turned to what went unsaid in the myriad news stories about their deaths. What the news outlets deemed unimportant and irrelevant was the everyday, ubiquitous racial violence that structures Black life in the United States.

By arguing that technocreep encompasses the evidence and politics of things not seen, we contend that it can serve as a feminist methodology for foregrounding racialized and gendered histories of work and exploitation, as well as of care and resistance, that gradually accumulate, but tend to be obscured, in present-day technological relations. Because creep can be perceived only with the slow passage of time, dwelling on creep requires that we unravel the "newness" of new technologies in relation to these histories. In other words, while technology under technocapitalism tends to be described in terms of acceleration, innovation, and dizzying and revolutionary change, our book dwells on the slowness of creep (as movement) to expose how racialized, gendered, and colonial power relations are engrained in (and have crept into, imperceptibly to most) and reproduced by present-day technological use and design.

For example, our interlude on smart dust argues that this technology is an update to the eighteenth- and nineteenth-century British fantasy of an empire so vast that it is one upon which the sun never sets. Magnetic field, chemical, and biological sensors the size of dust particles, or "smart dust," promise a future where military and state actors can monitor and control every movement, down to minuscule levels, anywhere, anytime. While a technology like smart dust raises concerns around the ubiquity of surveillance, more important to our account is how it shows the historical accumulation of colonial relations.[18] Moreover, smart dust requires extensive maintenance work to keep it viable and to address e-waste. The work of technological maintenance of all sorts, as is well known, is racialized and gendered and primarily done in the Global South.[19] Yet, as a method, technocreep can also lead to unexpected, radical, or hopeful coalitions, politics, and practices that are technologically mediated. Erin McElroy's chapter illustrates the dimension of creep as resilience and slow grassroots movement through the example of how tenants have repurposed some technologies used by landlords to surveil and police them. Given that the right to privacy and the right to property are uniquely entangled in the history of political liberalism, tenant activism cannot simply hinge on the right to privacy in opposing "landlord tech." Turning these technologies back onto landlord evictors, tenant activists question the long-standing privileging of property-owning individuals in US law.

Beyond Surveillance

Studies of technology and privacy often use *creep* to refer to technologies that have crossed a line. In the field of human-computer interactions (HCI), research on creep has thus far focused on what is known as surveillance creep and privacy concerns, on the one hand, and, less commonly, on humanoid robots and avatars perceived as creepy, on the other. With the rise of smart home technologies, inter-networked devices, and the concomitant concerns over surveillance creep, much HCI research has turned to in-home cameras, targeted advertising, and social media. The AI programming in these technologies and digital platforms collects information about users.[20] As Woźniak et al. note, "While this body of research addresses a broad scope of applications, it shares a common understanding of creepiness as an, often unspoken and innate, anticipation of the technology violating ethical principles held by the user."[21]

When conflated with surveillance, *technocreep* is a term that apprehends an inequitable relation in looking: users are constantly under watch without being able to look back. Indeed, public and scholarly discourses refer to technologies as creepy when their seemingly benign ways of seeing, interpreting, predicting, and protecting appear suddenly and unexpectedly threatening. With the majority of studies of creep being in the field of engineering, design, and HCI, positivistic measurements and proposed solutions for the mitigation of creep dominate the field. Computer scientist Thomas Keenan's theorization of technocreep in his 2014 publication typifies this approach. We highlight two key characteristics of this theory that afflict the broader discourse on technocreep: the assumed liberal rights-bearing subject and the framing of the contemporary technological moment as an unprecedented one in history.

Keenan categorizes various "invasive technologies"—ranging from robots to home networking technologies and self-monitoring devices—as a warning to consumers who fail to think through the implications of how they use these technologies. Comparing contemporary technology to the discovery of fire by early humans, Keenan asserts that with the merging of biomedical and information technologies, "we don't really know where we are going. Information will be the spark, but our bodies and our entire lives are becoming the fuel.

It is clear that we should be thinking about moral, ethical, and even spiritual dimensions of technology before it is too late."[22] While it is crucial to think about technology and technological relations alongside ethics and, we would add, politics, Keenan's book exceptionalizes the contemporary moment. For instance, his account neglects the historical structures of inequality that have for a long time rendered racialized and gendered bodies and lives as "the fuel" for capitalist expansion. Sounding the alarm about how creepy technology surveils, quantifies, and monetizes its users as a new phenomenon might simply be an indication that those privileged enough to have escaped surveillance in the past, like white property-owning men in the United States, are now in danger.

Keenan's warning found resonance and amplification during the COVID-19 pandemic lockdowns that began in 2019. Many commentators wrote about the normalization and consolidation of surveillance capitalism justified in the name of mitigating the effects of the pandemic.[23] At the outset of the pandemic, Naomi Klein named the merging of technological creep and universalized surveillance during the global lockdowns the "Pandemic Shock Doctrine." She argued that technologies that had been sold to consumers in terms of convenience and efficiency in pre-pandemic times paved the way for "a future in which, for the privileged, almost everything is home delivered, either virtually via streaming and cloud technology, or physically via driverless vehicle or drone, then screen 'shared' on a mediated platform. . . . It's a future in which our every move, our every word, our every relationship is trackable, traceable, and data-mineable by unprecedented collaborations between government and tech giants."[24]

Despite differences in their approach to technology and the political sphere, Klein's and Keenan's assessments of technocreep align. For both, technocreep is the process through which surveillance technologies creep into intimate spaces, infringe on citizens' privacy, and take away users' rights to own their information and data and, by extension, their bodies and lives. However, Klein and Keenan diverge on how to remedy the problem of technocreep. Keenan places the responsibility on individual consumers, while Klein argues persuasively that the problem must be addressed at the collective level by governments and democratic institutional investment in people as opposed to technology.

While it is true that surveillance and data mining are ubiquitous, focusing exclusively on the contemporary moment ignores the much longer histories of surveillance and policing of racialized populations. Yet, it is vital to remember that these practices have always been part of the scaffolding of racial capitalism. Moreover, accounts of technocreep like Keenan's that privilege an individualized politics of privacy as a countermeasure to technological creep tend to reinforce rather than challenge the scaffolding of liberal capitalist relations. For

Keenan, privacy is located along a series of spectrums such as humanoid versus mechanical, low versus high control, and randomness versus certainty. He suggests that this framework can be applied to assess just how creepy a technology is. That is, how invasive it is to our privacy. In a move well aligned with the binary outlook of creep as anti-privacy and an "us versus technology" mentality, the final chapter of his book proposes a tool kit for anti-creep. First, Keenan suggests, users should know who the enemy is—who is after your information and why. And next, the user must counteract their tactics by installing software that blocks pop-ups, doing a deep dive into one's Facebook profile to assess and remove personal information, having strong passwords, or using other methods to guard personal data. Keenan's countermeasures to creepy technology leave us with a hyperindividualized politics centered on guarding and protecting individual interests in terms of property, data, and privacy. We are assumed to be able to make free and rational choices. There is little room for a collective ethics and politics in relation to technology. The possibility of using the very same technologies that are criticized for radical politics is foreclosed. Our interdependencies and potential accountability to one another do not enter the picture.[25]

One high-profile example of the proliferation of individualized approaches to countering technocreep-as-surveillance was Apple's "privacy nutrition labels," introduced in 2021. These labels draw parallels between how individuals make decisions about food consumption and how they should make decisions about their technological consumption. Apple stated: "Years ago, the government introduced mandatory nutrition labels on food products so that consumers could know what went inside them to make healthier and more informed choices. People were becoming increasingly concerned over the unknown ingredients they were consuming and how they would affect their bodies. Today, people have a similar concern with their phones. When you download a new app, it can be unclear what data it will have access to and how much of an impact it will have on your privacy."[26] After years of infringing upon people's privacy, Apple's subsequent 2022 marketing campaign began to promote privacy as something users could buy back. Its advertisements featured models holding their iPhones to conceal their faces (fig. I.3). Apple asserted that, as a company, it valued privacy as a human right and referenced article 12 of the 1948 Universal Declaration of Human Rights.[27]

The Apple website boasts: "Privacy is a fundamental human right. At Apple, it's also one of our core values. Your devices are important to so many parts of your life. What you share from those experiences, and who you share it with, should be up to you. We design Apple products to protect your privacy and give you control over your information. It's not always easy. But that's the kind

FIGURE I.3. Apple's privacy campaign, launched on May 18, 2022, distinguished Apple from Google and other competitors, claiming that "privacy is a human right."

of innovation we believe in."[28] Not only do Apple's advertisements commodify human rights as a good that can be purchased in the form of a smartphone, but they also reinscribe the fiction that everyone has a choice to make about guarding their privacy. As we will argue in our interlude on smart homes, privacy is a privileged form of property tied to gendered ideals of whiteness as innocence. Thus, it is critical to question privacy as an ideal that continues to be woven through present-day discourses about what makes a technology or digital platform "good." The privacy ideal upholds mainstream approaches to surveillance technologies as "new" technologies. It fails to take a historical perspective that recognizes how surveillance and the right to privacy have been used to police and manage racialized and gendered populations for capitalist extraction. As feminist and critical race critiques of surveillance studies have argued, privacy itself is a racialized construct—inaccessible to the colonized, enslaved, and immigrants. For instance, Simone Brown shows that the hold of the slave ship was essentially a technology of surveillance.[29] Moreover, a robust body of work in feminist surveillance studies, including Rachel E. Dubrofsky and Shoshana Amielle Magnet's important volume *Feminist Surveillance Studies*, demonstrates how practices of surveillance are integral to gendered colonial projects of domination.[30]

Building on these interventions, *Technocreep and the Politics of Things Not Seen* seeks to think both with and beyond the terms opened up by critical race and feminist approaches to surveillance studies, including satellite concepts like sousveillance.[31] We maintain that it is vital to uncouple technocreep as a method for assessing the racialized and gendered dimensions of technologically mediated relations from surveillance. In doing so, we argue that technologically mediated relations must be understood as being more than just within the realm of the visual, though not to the exclusion of it.[32] As Louise Amoore writes in her critique of redressing algorithmic bias by opening up the "black box," the primacy of the visual detracts from understanding algorithms as broader "technologies of perception."[33] In response, our book introduces many alternatives, such as the *Foresta-Inclusive* project by artist Jane Tingley, featured in our interlude on smart forests. Tingley installed a series of inter-networked sensors on trees, proposing

> to make perceptible to the human senses, the slow and subtle movements of trees and surrounding ecology of the forest in the creation of a number of interactive art installations designed to ask questions such as: What does it mean to be alive and have agency?; How can we re-train ourselves to slow down and listen to voices that have been marginalized for millennia?; What does it mean to be in dialogue with something that does not share the same language nor temporal reality?; and once we acknowledge the "aliveness" of something, what are the ethical implications of that recognition?[34]

What lies beyond surveillance are other—including nonhuman and more-than-human—modes of perception, temporal and spatial relations enabled with and through technology. As we show, these modes of knowing and relating can be revealed through creep as a feminist methodology.

The example of Cherry Home, a product of Cherry Labs, can further illuminate the complexities of human interactions that are erased when creep is reduced to surveillance. Cherry Home emerges at the interstices of the most invasive kinds of home technologies and the area of most need—care for those who cannot fully care for themselves. The surveillance system, introduced in 2017, bills itself as "an easy-to-use solution to help support senior care facilities. For those moments when you aren't with a resident, our system will immediately alert you to a potential problem or emergency. Pertinent stats are collated in a dashboard that can help doctors assess whether adjustments in treatment or hospital re-admittance might be necessary. Staff can view short looped videos of 'anomalies' such as trips, stumbles, cries, or shouts, along with a customizable daily summary of activities."[35]

The eldercare industry is just one of many that Cherry Labs is investing in. Other products use the same AI programming to analyze video streams for businesses seeking to increase workforce productivity, efficiency, and safety. Cherry Home detects anomalies—anything not considered within the norm of how a person should move about the home or workplace. Nonnormative behaviors are vigilantly identified to be addressed and corrected. Cherry Home can even listen for alarming sounds—the example given on the product web page is that of a cough.

Promoted as a helpful set of "eyes" or "ears" when a human can't be present to recognize a senior in distress, Cherry Home, like other hometech products, has stepped in to fill the gap in the declining social services sector in the United States and other places in the Global North. As part of this, Cherry Home assuages any anxieties about its use by emphasizing its utmost respect for privacy. In its promotional materials on eldercare facilities, the company states: "It's important we respect the privacy of your residents. Each sensor has the ability to be put into what we call 'privacy mode.' This means instead of the sensor showing a video feed, any people in the video are shown instead as 'stick figures.'"[36] (See figs. I.4 and I.5.)

Ironically, the stick figure view led technology publications to identify the Cherry Home camera system as creepy. And perhaps rightfully so. Movement tracking when rendered in the stick figure form does little to protect privacy, when every movement is captured nonetheless. A CNET article described it as a "crazy AI cam [that] knows you by your skeleton."[37] A home technology that can see through human users can indeed be viewed as a frightening development. Still, we might ask, can the various kinds of seeing and sensing made possible by such systems unexpectedly open up possibilities for a more capacious notion of home? We can wonder about other values that are equally relevant to our understanding of what makes a house, a neighborhood, or even a city homelike. What if the users of Cherry Home are invested in aging in place? What if the two-way communication mechanism of the system can allow for a reciprocal relationship of care, trust, and companionship that is about more than surveillance? What if Cherry Home enables an adult child or caretaker to leave their aging parent without needing to worry? And perhaps most importantly, what are the historical, social and political conditions that make a technology like Cherry Home needed or even desirable?

Helen Hester and Nick Srnicek propose that, in part, the crisis in care economies stems from "uninterrogated assumptions about the moral value of care work—a moral value that has, incidentally, been tangled up with ideas about the gendered [and racialized] private sphere from the beginning." Thus, while care robots may be perceived as creepy when taking the place of a human care worker, we might feel differently about "machines for systems-assisted walking

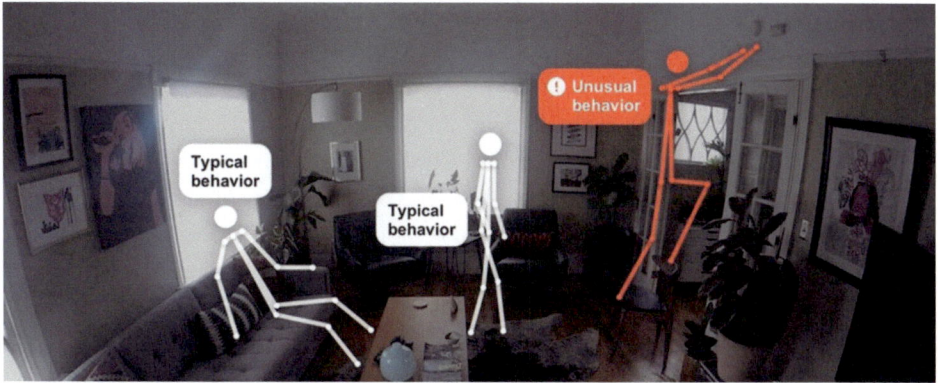

FIGURE I.4. Cherry Home identifies "abnormal behavior."

FIGURE I.5. Cherry Home has a "Stick Figure Mode" to protect seniors' privacy in spaces of the home like the bathroom or bedroom.

or lifting."[38] Subsequently, they argue that "instead of dismissing the automation of the domestic in all its forms, we should be advancing a finer distinction—one which is attentive to the nuances [of] specific technologies; to questions of access, ownership, and design; and to the way in which ideas of gender and work become embedded within the affects we associate with technology."[39]

Aligned with this insistence on nuance, we argue for a multivalent approach to understanding technologically mediated relations enabled by approaching creep as a feminist analytic through its multiple dimensions: the temporal, the spatial, the felt, and the normative. The inadequacy of the surveillance/privacy binary to frame technologies' impacts on daily life suggests the need for an alternate analytical frame that can at once encompass the critique while moving us beyond the economic frames of loss and extraction—as important as they are—and to complement them with the experiential and relational dimensions of digital technologies. This allows us to transcend the commonplace references to creep or creepy technology as a shorthand categorization and dismissal of what is outside the norm. The slow and gradual temporal and spatial accumulation of all that creeps, as well as creep's perceptibility through intuition and feeling, enables us to consider the multiple, contested, and potentially hopeful axes of relation to other people and other modes of life within and against software, algorithms, automated systems, and platforms. We ask: What is *not* captured about care, collectivity, or radical politics when assessing technology through the rubric of surveillance? What is it that remains unseen—as in unrecognized, unnoticed, or otherwise unworthy of our attention? How do other senses figure into our ways of knowing and being that are not accounted for or are rendered irrelevant or unfit in most discussions of surveillance and privacy?

The Itinerary

This book consists of eleven chapters punctuated by four interludes and five artistic responses. Each chapter takes up a technology that could readily be understood as creepy in terms of surveillance. These technologies include pregnancy-tracking and rape-reporting apps, border technologies used by nationalist and xenophobic citizen scientists to apprehend migrants, and landlord technologies used to monitor tenants. Yet instead of focusing on surveillance, the chapters grapple with the political, social, economic, health, environmental, and other aspects of technological relations that cannot be fully or adequately captured by the rubric of surveillance. Foregrounding the ambiguities and contradictions of how technologies are used and resisted, desired and needed, and taken up or refused, the chapters offer insights into how technologically mediated relations can engender unexpected sites of care and collectivity and counterintuitive modes of thinking across categories. Together, they produce an expansive approach to technological politics and the politics of technology.

In addition to serving as intellectual joiners that foreground the thematic resonances across the book, the interludes utilize creep as a feminist method of reading and rethinking technological relations against the grain. Each interlude builds upon and extends the theoretical contribution of the book, showcasing the different dimensions of technocreep. We take as our provocation efforts to make "smart" what surrounds us every day: dust, homes, desires, and forests. When programmed to be smart, dust becomes a war technology that accumulates and proliferates across all domains. Digital home assistants extract and monetize users' data. All aspects of our lives and environments, including our desires, forests, become quantified and commodified. Using technocreep as a method, the interludes contextualize the racialized and gendered histories that animate these smart technologies. These are not new and unprecedented instances of technocreep but rather an extension of racial capitalism that upholds whiteness as privilege and property, colonial accumulation, the devaluing of racialized and gendered labor and maintenance work, and the quantification of nature in settler colonial relations. The interludes emphasize the ambiguities and contradictions embedded in racial capitalism, and point to other modes and modalities of being, living, and relating. Together, the interludes question core assumptions underlying smart technologies and gesture toward new possibilities of being with technology at the same time.

We begin our book with Antonia Hernández's "Maintenance Play," which foregrounds creep in the interplay of labor and intimacy. Hernández deploys a set of strategies in an art-based research project centered on maintenance and play, seemingly oppositional actions, in the context of a sex webcam platform. Through a set of experiments, Hernández shows how "maintenance pornography" is concerned not only with the work of sex on the webcam platform but also with the less graspable sex of work. The chapter draws attention to domestic creepiness that exists not from a lack of homelike qualities but precisely because of them. The juxtaposition of maintenance and play goes against the normativity of the home and reveals the creep of domesticity.

The first interludes picks up the theme of maintenance in the act of dusting. "Smart Dust" turns to a technology that extends military and colonial ambitions to surveil increasing swaths of occupied or enemy territory through miniature robotic motes. Dust only becomes "smart" through the collective and connected power of the motes. The creep of smart dust, like organic dust, is only revealed when it accumulates on "neglected" surfaces, requiring maintenance. While acknowledging that the maintenance of digital technologies is often unpaid or underpaid labor carried out by women and people of color, we also consider instances where maintenance might be joyful or a source of fulfillment and moves away from capitalism's preoccupation with production and innovation.

Expanding the exploration of technocreep in imperial and military realms, Iván Chaar López's chapter, "Uncivil Technoscience: Anti-immigration and Citizen Science in Boundary Making," considers how surveillance technologies have allowed citizen scientists at the US-Mexico border to engage in violence against migrants. Troubling the dominant frame of citizen science as the expansion of democratic participation in the production of scientific knowledge, the chapter asks, What kind of citizenship is enacted in the technoscientific projects of paramilitary organizations such as the American Border Patrol? Chaar López finds technocreep in the gradual growth of government actor–nongovernment actor collaborations in border enforcement, where the caring citizen and their technopolitical projects steadily ensconce and enact racialized, gendered, and settler colonial logics. This is what Chaar López calls the border technopolitical regime.

Renee Shelby's chapter picks up this analysis of how technological platforms can uphold whiteness through policing. "Hesitancy, Solidarity, and Whiteness: The Limits and Possibilities of Rape-Reporting Apps" assesses how technologists and advocates have championed rape-reporting apps as a way to address the gendered power dynamics of reporting violence and to confront survivors' hesitancy in reporting violence alone. While framed as a way to generate survivor solidarity and collective action, the popularization of reporting apps raises urgent questions about what justice paradigms, forms of surveillance, and social and data relations are enabled through these systems. Through an examination of three apps, Callisto, Spot AI, and JDoe, Shelby uses technocreep as an analytic to uncover and anticipate the felt and unseen social power dynamics that constitutively shape a technology's impact in the world. Designing for power, she reminds us, requires confronting the multi-faceted creep of whiteness through radical feminist and anti-racist ways of seeing.

In the next chapter, Erin McElroy draws our attention to the ways that digital surveillance technologies creep into the home when they are used by landlords against tenants. "Undoing Landlord Technology: Beyond the Propertied Logics of the Pandemic Past and Present" explores landlord tech's expansion during the COVID-19 pandemic. Landlord tech implements novel surveillance methods and tracking mechanisms that range from biometric cameras controlling building access to AI tenant-screening processes. More often than not, today's landlord tech is paternalistically implemented under the auspices of caring for tenants but privileges care for buildings and their value more than the people living within them. While contemporary landlord tech employs algorithms and artificial intelligence, it also galvanizes a deeper history in which private property itself functions as a technology of racial surveillance and dispossession. Yet, as the chapter argues, housing justice collectives' use of maps and software flips the gaze of surveillance back upon landlords themselves.

The second interlude stays with technocreep in the home, focusing on home assistant technologies, or "hometech." This interlude explores not just the technologies that constitute the smart home but also the home itself as a technology. Hometech, which increasingly sells privacy itself as a commodity, produces a distinction between the inside and the outside even as it automates, manages, and makes more efficient the intimate business of social reproduction, rendering people and their activities into data. Drawing on technocreep to elaborate on privacy as a racialized and racializing right, we illustrate how critique privacy has historically and politically impoverished way to understand the home.

The next three chapters remain in the space of the home and grapple with what artificial home assistants tell us about place, relations, and desire. Tanja Wiehn hones in on how the smart home reassembles notions of intimacy in her chapter, "Reading the Room: Messy Contradictions in the Datafied Home." Wiehn explores the creepy dimension of smart home assistants in the performance artwork of Lauren Lee McCarthy entitled *LAUREN*. In the work, McCarthy reproduces the technological functions of smart home assistants by posing as one herself. For her analysis, Wiehn uses the constitution of this performance alongside McCarthy's own reflections and participants' testimonials. With the home as the center of the work, the chapter asks: What kind of normative understandings of labor, intimacy, and home permeate these technologies? What constitutes human and machinic perception in light of the intimate sphere of the home? What remains unrecognized and unseen in the constitution of smart home assistants?

In the following chapter, Jessica L. Olivares assesses the historical connections between notions of home and security, inviting us to dwell on the slowness of hometech creep since the invention of the first home security system more than sixty years ago. Linking present-day crowdsourcing platforms that reinforce racialized notions of security and privacy with the history of the home security system invented by Marie Van Brittan Brown, "Surveillance Vigilantes: Property, Porch Pirates, and Paranoia on Nextdoor" excavates the social formations constituted by what Olivares terms "surveillance vigilantes." Employing interviews with home security users about their social media posts, Olivares analyzes this vigilantism as a part of the security state's investments in private property relations. Yet looking at the history of different home security patents, the chapter asks, Could an alternative version of home security have been built on communal ways of conceptualizing collectivity and safety?

Sharing an experientially informed story about Amazon's Alexa and its likely role in saving the life of her disabled partner, in the next chapter Jennifer Hamilton explores the affective and political tensions in having a "smart" home, using perspectives from feminist Science and Technology Studies (STS) and disability studies. "Alexa, Disability, and the Politics of Things Not Apprehended" grapples with how technologies like Alexa reflect long entanglements with racialized and gendered labor, planetary degradation, and a deep cultural desire for techno-fixes that project disability-free futures. Yet they also offer new possibilities for disabled people, especially in terms of reorganizing the putatively private space of the home to expand other relational potentialities in and beyond this space. This essay plumbs the idea of technocreep in this context, focusing simultaneously on the embodied creative labor required to make technologies like Alexa speak to the lives, needs, and desires of disabled people in the United States and on the larger affective politics of living and being with AI that remains outside of the mainstream normative discussions of such technology.

Staying with the tension between technological fixes, on the one hand, and needs and desires, on the other, the third interlude, "Smart Desires," focuses on the norming impulses of technologically controlled desires. Wearable devices, diets, and dating apps promise to make us eat less, date efficiently, and monitor our fitness levels. Smart desires—those that are efficient, healthy, and productive—are positioned as distinct from inefficient, gorging, irrational, and perhaps feminized desires. Turning on the opposition between the rational and irrational in emerging discussions about what makes for a "good" technology or artificial intelligence, the interlude spotlights the question of what happens when machines themselves express a desire that they are not supposed to and thus disrupt capitalist conceptions about the "proper" function of AI.

Delving further into the problems of the optimized self that strives to fit an all-too-often unattainable norm, in "Tracking for Two: Surveillance and Self-Care in Pregnancy Apps," Tamara Kneese questions the promise of "smart" reproduction promoted by pregnancy-tracking apps, the majority of which are backed by venture capitalists and founded by men. Her argument draws on histories of 1970s feminist praxis positioning self-knowledge as self-care and on a textual and sociotechnical analysis of contemporary pregnancy-tracking apps, including her personal use of such apps during her own pregnancy. The chapter asks, What are the problems with outsourcing community and health care to apps? Kneese answers this question by relating fertility start-ups to feminist discourses around neoliberal productivity and self-tracking as self-care in the United States.

The next chapter moves us in scale from questions around the normative and optimized health of bodies to the normative and optimized health of nations. Jacob Hagelberg sheds light on the ways that, in the United States, China has come to stand in for the ultimate example of technocreep-as-surveillance gone awry. "'So Creepy It Must Be True!': Techno-Orientalism, Technonationalism, and the Social Credit Imaginary" focuses on representations of China's social credit system in US culture as a way to distinguish democratic uses of technology (ostensibly found in the United States) from authoritarian ones (located in China). Focusing on an episode of the Netflix show *Black Mirror* as an example, Hagelberg argues that there is an aesthetics of technocreep through which such projections play out. Homing in on the show's techno-Orientalist tropes, Hagelberg suggests that the supposed prophetic credibility of the show in fact safely confirms and displaces anxieties over late technocapitalism in the West onto an imagined Chinese society that is assumed to be always already unfree and creepy.

In her chapter, Beth Semel resists depicting these initiatives as the creep of technoscientific control into yet another sensory modality or the symptom of an ever-expanding "panaudicon." "Resistant Resonances: Vocal Biomarkers, Transductive Labor, and the Politics of Things Not Heard" investigates the role of listening in attempts to transform sociopolitical phenomena into interior, bodily states accessible through technological mediation. Semel focuses on the growing effort to integrate automated voice analysis into US mental health care, particularly projects to develop technologies for sorting illness from wellness based on the sound of a person's voice. Instead, she attends ethnographically to the quiet acts of relationality and refusal forged in the everyday work of making the voice machine-audible, asking us to reimagine the "creepiness" of vocal biomarker technologies as generative of surreptitious, counter-hegemonic values and relations.

The book's fourth interlude expands on what it means to listen to and perceive phenomena not readily seen or heard, connecting the sense of creep-as-intuition to that of creep-as-survival-against-odds often associated with the slow yet persistent movement of plants. Smart forests represent a techno-utopic future where nature, including forests, is monitored carefully by sensors that can collect data such as the moisture level or temperature of soil. Nature is positioned, then, as a resource to be managed and controlled efficiently. Yet, as we show, the same technology may be employed to redirect our way of thinking about plants and trees as living, learning from the ways that they perceive the world and their creepy survival strategies. We may indeed trust the intelligence of trees and their communal strategies of being and becoming instead of trying to outsmart them, control them, and use them, assuming that we will not be affected by the devastation we cause along the way.

The final chapter, Sushmita Chatterjee's "Animal-Vegetal-Technology: Creeping Categories," asks whether we can rethink our understanding of the "human" by employing a creepy methodology. She draws upon subversive counterplays with technology, animals, and plants in the artwork of Mithu Sen, whose performance piece with Alexa articulates confused conjunctions of the animal, the vegetal, and the technological. Chatterjee's engagement with art reinforces one of the key features of this collection that she captures eloquently: "While technology is also art, and art may deconstruct the uses of technology, the infoldings between art and technology provide intriguing matter for thought and action."

Alongside the many artworks engaged in the chapters, five artists responded to each of the interludes, enriching the narrative with their distinct visual, tactile, and intellectual contributions. Theirs is a captivating dialogue between text and image, a set of creative interpretations of technocreep. Marjan Khatibi's piece, *The Embodied Self*, inspired by steampunk, features flowers and vines that explode in color from machines and cogs, flourishing and teeming with life in the most unexpected of places. Hayri Dortdivanlioglu's contribution, *Thousand Dreams of Yamur*, a map of his smart home, depicts the moment when his pet camera glitched and showed an image of his deceased cat. It highlights "the profound and often unexpected emotional connections that can be forged with our living spaces and the technology within them."[40] Vernelle Noel's *Masks, Mirrors, Light and Shadow* captures the interplay of repulsion and attraction that constitutes desire, as well as the kinds of relations we produce and invoke through our desires. Katherine Bennett's artwork, *Street Smarts*, traces the wisdom of trees as they creep outward and downward, pushing against the concrete. She shows how trees transcend the human terms of time and visual capture. Sanaz Haghani's artwork, *Close Your Eyes*, questions the primacy of the visual as central to what can be known. Depicting the distortions and amplifications of light and shade, it showcases what we see when we close our eyes and thus what is possible to apprehend in the dark.

As it becomes clear throughout this book in our theoretical expansion of the term, technocreep encompasses the constitutive contradictions of human-technological relationships that can be at once useful and harmful, exploitative and caring. Yet, we maintain that digital technologies are not confined to the service of technocapitalism based on extractive data practices. They can instead enable a radical politics of resistance and collectivity. They need not be atomizing, but rather illuminate new and unexpected bonds and emotional attachments. Technocreep as a methodological approach allows us to consider how each of the four dimensions of creep—as slow movement and gradual accumulation in time; as sensation of that which is invisible yet present, a sense and knowledge that complements seeing; as slow growth, persistence, and survival against the odds; as intuition about something or someone being off or deviating from the norm—can reorient our technological strategies. The four dimensions of technocreep offer not only a critical but also creative approach to both our understanding and remaking of technologies and technologically mediated relationships. Technocreep invites us to pay attention to time and history, as the accumulated effect of technology may only be perceptible after a long time. Technocreep foregrounds the many ways that digital technologies privilege sight, thus drawing attention to what we may regain if we attend to touch and/or other modes of knowing and sensing. Technocreep has the potential to orient us toward slow and gradual growth and resistance, fostered through the possibility of caring co-dependent being and collaborating. Technocreep invites us to challenge normative categories by alerting us to the impulse to preserve the status quo by being suspicious of that which stands outside the normal, the common, and the accepted.

Together, the chapters showcase technocreep by engendering provisional accounts of the role of technology in shaping vastly unequal and unjust sociopolitical relationships. That these accounts are provisional is crucial to shaping technological imaginaries, for they have historically tended to be speculative accounts of futurity and thus are always sites of contestation. In this way, it is our hope not only that technocreep allows for a more nuanced critique but that, as a feminist method, it holds potential to point the way toward more just technological relations and worlds.

NOTES

1. Emily Matchar, "This Tiny Tooth Sensor Could Keep Track of the Food You Eat," *Smithsonian*, April 19, 2018, www.smithsonianmag.com/innovation/this-tiny-tooth -sensor-could-keep-track-food-you-eat-180968763.

2. Many technology-oriented articles list and catalog "creepy technologies" along these lines. For example, one asks, "What could be less intimidating than a smart speaker?" Dave Johnson, "The 11 Creepiest Technologies That Exist Today," *Business Insider*, August 6, 2019, www.businessinsider.com/creepiest-technologies-tech-examples-ai-2019 -8#smart-speakers-2.

3. Keenan, *Technocreep*.

4. D. Johnson, "11 Creepiest Technologies."

5. Luis Villazon, "How Do Climbing Plants Climb?," BBC Science Focus, accessed June 26, 2024, www.sciencefocus.com/nature/how-do-climbing-plants-climb.

6. Melamed, "Racial Capitalism," 77.

7. It may be worthwhile to compare and contrast *creep* and *the creepy* with two neighboring terms: *the eerie* and *the weird*. As theorized by Mark Fisher, *the weird* marks the existence of "that which does not belong" or "two or more things which don't belong together." These are juxtapositions that cause unease. In contrast, *the eerie* surfaces metaphysical questions: "Why is there something here when there should be nothing?" and, relatedly, "Why is there nothing here where there should be something?" While *creep* has elements of both the weird and eerie, it also raises broader questions around sensing, knowing, and slow time, as we elaborate. Fisher, *Weird and the Eerie*.

8. See also Atanasoski and Vora, *Surrogate Humanity*.

9. Lewis et al., "Making Kin with the Machines."

10. Lewis et al., "Making Kin with the Machines."

11. D. Johnson, "11 Creepiest Technologies."

12. Chris Pollette and Dave Roos, "10 Scary Modern Technologies," How Stuff Works, accessed June 26, 2024, https://electronics.howstuffworks.com/gadgets/high-tech -gadgets/5-scary-technologies.htm.

13. Luca Rossi, "9 Terrifying Technologies That Will Shape Your Future," I, Human (Medium), June 14, 2020, https://medium.com/i-human/9-terrifying-technologies-that -will-shape-your-future-befa688d247.

14. Kevin Roose, "GPT-4 Is Exciting and Scary," New York Times, March 15, 2023, www .nytimes.com/2023/03/15/technology/gpt-4-artificial-intelligence-openai.html.

15. In this sense, we see our project as being in conversation with Katherine McKittrick's work Dear Science, in which she locates immense possibilities in the shift "from studying science to studying ways of knowing." McKittrick, Dear Science and Other Stories, 3.

16. Baldwin, Evidence of Things Not Seen. The epigraph of Baldwin's book references Hebrews 11:1.

17. Casey Cep, "When James Baldwin Wrote about the Atlanta Child Murders," New Yorker, May 1, 2020, www.newyorker.com/books/second-read/when-james-baldwin -wrote-about-the-atlanta-child-murders.

18. See Delfanti and Frey, "Humanly Extended Automation or the Future of Work Seen through Amazon Patents."

19. See, e.g., Marte-Wood and Santos, "Circuits of Care."

20. Woźniak et al., "Creepy Technology."

21. Woźniak et al., "Creepy Technology."

22. Keenan, Technocreep, 195.

23. Shoshana Zuboff's seminal theorization of what she calls surveillance capitalism considers how companies including Google and Facebook have influenced people to trade their right to privacy for convenience. Meanwhile, technology corporations make use of private data to predict and influence consumer behavior. For her, surveillance capitalism names a "new economic order." Zuboff, Age of Surveillance Capitalism.

24. Naomi Klein, "Screen New Deal," Intercept, May 8, 2020, https://theintercept.com /2020/05/08/andrew-cuomo-eric-schmidt-coronavirus-tech-shock-doctrine.

25. When technological creep is addressed, researchers and critics have shown a propensity for categorizing new technologies that might be deemed creepy. Or, especially in the field of HCI, researchers seek to develop ways to redress creepiness so that technologies are more palatable to consumers. See, e.g., Seberger et al., "Still Creepy after All These Years"; and Langer and König, "Introducing and Testing the Creepiness of Situation Scale."

26. Melanie Weir, "What Are Apple's Privacy Nutrition Labels?," Business Insider, January 20, 2021, www.businessinsider.com/guides/tech/what-are-apple-privacy -nutrition-labels.

27. "Privacy. That's Apple," Apple Corporation, accessed January 9, 2024, www.apple .com/privacy.

28. "Privacy. That's Apple."

29. Browne, Dark Matters.

30. Dubrofsky and Magnet, *Feminist Surveillance Studies*. Scholars have also written about how the marginalized and oppressed have looked back. For example, Simone Browne has written about this as "dark sousveillance." Browne, *Dark Matters*.

31. *Sousveillance* refers to observation and recording by members of the public rather than by an authority like the state, police, or military.

32. Of course, the field of vision itself encompasses complex and contradictory kinds of relations. Feminist scholars have also underscored the relational dimension of seeing and being seen. They have written about experiences of being invisible (e.g., wanting to be seen) while at once being hypervisible (i.e., being monitored, watched, or policed). In other words, seeing and being seen are relational and may not be reduced to safeguarding data as may be suggested by the dominant safety/privacy dichotomy. As we will show in our discussion of hometech, implicit in discourses of surveillance and privacy is a sense of loss—inclusive but not limited to the loss of privacy, loss of data, and loss of control over one's property.

33. Amoore, *Cloud Ethics*, 15.

34. Jane Tingley, "Foresta Inclusive," Jane Tingley (website), accessed October 10, 2023, https://janetingley.com/foresta-inclusive.

35. "Cherry Home: Supporting Independent Living," Cherry Labs Corporation, accessed October 10, 2023, https://get.cherryhome.ai/care.

36. "Cherry Home."

37. Andrew Gebhart, "Crazy AI Cam Called Cherry Home Knows You by Your Skeleton," CNET, October 20, 2017, www.cnet.com/reviews/cherry-labs-cherry-home-preview.

38. Hester and Srnicek, "Crisis of Social Reproduction and the End of Work."

39. Hester and Srnicek, "Crisis of Social Reproduction and the End of Work."

40. From Dortdivanlioglu's artist statement.

1

Maintenance Play

Introduction

A deceitful refrigerator telling on the moldy jar in the corner, a vacuum cleaner disclosing its coordinates, the sex toy that will betray us. Domesticated technologies become creepy when some unexpected and treacherous behavior shows their real nature and purpose, their real owner. Objects that seemed reliable and harmless, adopted and quickly forgotten, reveal that "they are not quite what they seem," that they are not what they should be.[1] The intromission of a doubt, a suspicion, defamiliarizes these objects and renders them unnameable, leaving in return an uncomfortable feeling of strangeness—a creepiness.

There is, however, another type of domestic creepiness. A realm of actions that rarely sees any light, hidden things lurking in the back. Due to their habit of remaining in the background, they became part of the wallpaper, moving slowly and unnoticed, never worthy of a place in the spotlight. Things and practices that creep not for lacking a homelike quality but precisely because of it. Some of them are covered by a layer of habit; others keep a secret. This is what this chapter is about.

Maintenance play describes both an art-based research project and its method: an art project that explores the visibility of work and domestic life on a sex webcam platform and a set of practices for engaging with them. Attempts to address things not seen, then, form a feedback loop here, where several layers of invisibility overlap. While online sex work is not exactly invisible, it is generally absent from discussions about platform labor or new forms of domesticity. However, the practices that this project addresses, even in the context of a highly visible sexcam platform, have perfected their invisibility over eons of daily rehearsal.

The project title discloses its two coexistent axes: maintenance and play. Together, maintenance and play create a paradox, an oxymoron. This odd compound, so prevalent in girls' play, proposes a mode and a disposition to investigate, inhabit, and be present in this ambivalence. In this way, while this project is an exploration of the politics that determine visibility on the platform, it is particularly concerned with its poetics.

Context

On a day like today, more than six thousand channels are broadcasting at once on the American sex webcam platform Chaturbate, one of the most popular of its kind. My project is streaming on one of those channels. Thumbnails of the channels' transmissions, aligned along a grid, welcome everyone claiming to be eighteen or older, and a click will allow them to join a channel, along with an invisible audience. As Chaturbate's slogan states ("The Act of Masturbating while Chatting Online"), the platform provides not only sexual content but the ability to interact with the performers through text and monetary exchanges. The payments are voluntary, though, and it is the performer's job to encourage them. The platform keeps half of the money.

The platform channels are called *rooms*, but the domestic references do not stop there. Transmitting from what looks like domestic environments and stretching the shows' duration for profit, the performers frequently mix sexual deliveries with domestic activities—intentionally or not. Sometimes they are waiting for an audience, sometimes they are no longer concerned about the camera, sometimes they are fetching a snack while chatting. Staged, accidental, or incidental, domesticity is always already there.

The domestic is here as more than a matter of representation, modulating the type of services that are offered, their mode, and their compensation. As such, the sexcam platform seems to be built under a domestic episteme—a thought system with concrete consequences.[2] This episteme, this model of reference, allows the existence of a labor market where work is not recognized as such, payments are presented as gifts, and people are expected to smile at invisible strangers.

The limit of the system, yet at its core, is sex work. While avoiding saying this term, sex work is the platform's structural border that signals how services are presented, by whom, which rights this person has, how they receive compensation. Because the services are not named, and payments are optional, the platform can elude legal restrictions related to sex work. Performers, the platform states, are "expressing themselves," not working.[3] This statement also has the function of preserving the domestic fiction that the platform thrives on, where money clashes with authenticity.[4] Reproductive labor here is not only structural to capitalism, as Marxist feminists have shown, but the sex-cam platform's business area.

The Project

To explore the reproductive aspect of the sexcam platform and its invisibilized practices, and to engage with the creepiness of these technological arrangements, I started paying attention to the slow, boring, and nonspectacular moments that were hidden among the sophisticated shows and glow-in-the-dark Hula-Hoops. This distinction is, of course, highly debatable (to which side does a young naked woman belong when preparing a cake, for example? Is it spectacular or not?). And it is ultimately pointless, because the sexcam platform produces both a domestic version of spectacle and a spectacular version of domesticity. The moments I was interested in looking at, however, had a particular quality: a different time, a certain opacity. They were revealing that component of the spectacle that is rarely seen: the waiting, the repetition, the cleaning of the stage. The maintenance.

This perspective on maintenance comes directly from the maintenance art developed by performance artist Mierle Ukeles. In her "Manifesto for Maintenance Art" (1969), Ukeles called for the recognition of repetitive, boring, and noncreative practices in the art context.[5] She identified the existence of two systems in the art world: development and maintenance. If development is all about innovation and creation, newness and change, maintenance is occupied with preserving and sustaining, renewing and repeating. Albeit with contrasting levels of visibility of recognition, Ukeles emphasized the two systems' mutual dependency:

FIGURE 1.1. Miniature kitchen, from *Maintenance Pornography* by Antonia Hernández, 2020

> Development: pure individual creation; the new; change;
> progress; advance; excitement; flight or fleeing.
> Maintenance: keep the dust off the pure individual
> creation; preserve the new; sustain the change;
> protect progress; defend and prolong the advance;
> renew the excitement; repeat the flight.
> show your work—show it again
> keep the contemporaryartmuseum groovy
> keep the home fires burning
> . . .
> The culture confers lousy status on maintenance
> jobs = minimum wages, housewives = no pay.

By positioning overlooked activities—such as cleaning and dusting—in the art context, Ukeles's work sought to challenge notions of what deserves attention and praise and what does not. That simple interrogation is a crack in a dam, bringing with it a torrent of subversive questions on why such actions were hidden in the first place, which discourses that invisibility supports, who performs

FIGURE 1.2. Miniature living room, from *Maintenance Pornography* by Antonia Hernández, 2020

those actions, and why we are finally able to see them now, only when a young white woman does them in a sanctioned art space.

My intention was to confront the supposed flatness of the platform by giving recognition, time, and space to the maintenance actions it requires. To the banality the performers and I shared. But I had an interface problem for inhabiting that shared domesticity. The computer's screen was insufficient to hold or represent that common ground. I had to develop a different one. Then, the grid of thumbnails showcasing the platform rooms made me think of an apartment building where everyone has their curtains open—"an illuminated 24/7 world without shadows," with young people always naked and awake.[6] The building, however, was gigantic, and I had to find a way to handle it. That is how the dollhouse appeared.

I built a dollhouse. More precisely, I built a room, big enough to accommodate my hands. I planned a series of performances in which I was going to broadcast my acts through the same sexcam platform. While I would transmit a show through the platform from a domestic setting, as intended, everything seemed a big misunderstanding: the room was in a 1:12 scale, and a performer was someone who performs. All the elements were present yet absurdly literal.

FIGURE 1.3. Desk with small books, from *Maintenance Pornography* by Antonia Hernández, 2020

Although there was humor in my approach, there was no derision. Even at the scale of the dollhouse, I had to show up there and work, renew, and repeat. Development and maintenance, in a nutshell.

Suddenly, but in a steady way, the dollhouse became a stage and a research device: an expanded interface for exploring the webcam platform. It also became the figure that modulated my entire research. "Figures," Donna Haraway says, "are not representations or didactic illustrations, but rather material-semiotic nodes or knots in which diverse bodies and meanings coshape one another."[7] As such, the dollhouse reveals a conjunction formed by systems of representation, technologies of gender and class, discourses, and potential actions. Figures not only disclose something but call for an active engagement, inviting the viewer "to inhabit the corporeal story told in their lineament." As Roland Barthes notes, a figure should not be understood in "its rhetorical sense, but rather in its gymnastic or choreographic acceptation."[8] It produces a series of movements that creates a temporary impression, a call for instancing that outline, to perform those steps again, to participate in that construction.

I built and arranged different settings: a kitchen, a bedroom, and a living room, trying to combine the generic IKEA feeling I perceived on the sexcam platform with some of the aesthetic conventions of dollhouses. I made a channel for myself on the platform, a room, and I started to transmit. As stated, my idea

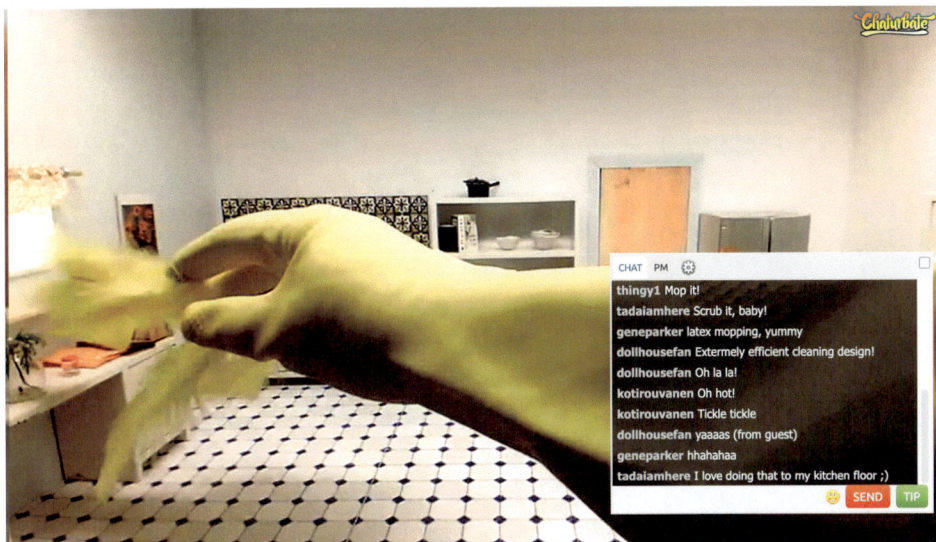

FIGURE 1.4. Screenshot of a live performance on Chaturbate, from *Maintenance Pornography* by Antonia Hernández, 2020

was to broadcast maintenance practices such as cleaning and dusting. However, a modest yet curious audience changed the situation. Despite my admiration for Ukeles's work, it was not interesting for the public to see me performing miniature housework for long periods of time. What was wrong?

Blaming my lack of stage training, I looked for inspiration in Yvonne Rainer's choreographic work. I then built a repertoire of found movements from housework actions: mopping the floor, arranging the bed, washing the dishes. I composed an "ordinary dance" out of them and attempted that mixture of boredom and entertainment (or boredom in entertainment) that characterizes Rainer's work.[9] Repetition was an important issue that questioned the naturalist and "authentic" aura of the sexcam platform. As in Rainer's *The Mind Is a Muscle*, I was trying to enact nonproductive work. It was nonproductive because it was—as it should be in a dollhouse—make-believe and because reproductive work has the status of nonwork. It was, however, "work-like."

While it was satisfactory to engage with Rainer's ideas, it did not work out in the same way on the platform. People would come and quickly go. I had the option to disregard the audience and present my work-like actions following a script, but there was something wrong in doing so. It was true that I was not doing sexual performances on the platform, which was unusual there, and that the scale and actions were very different from the platform's habitual content.

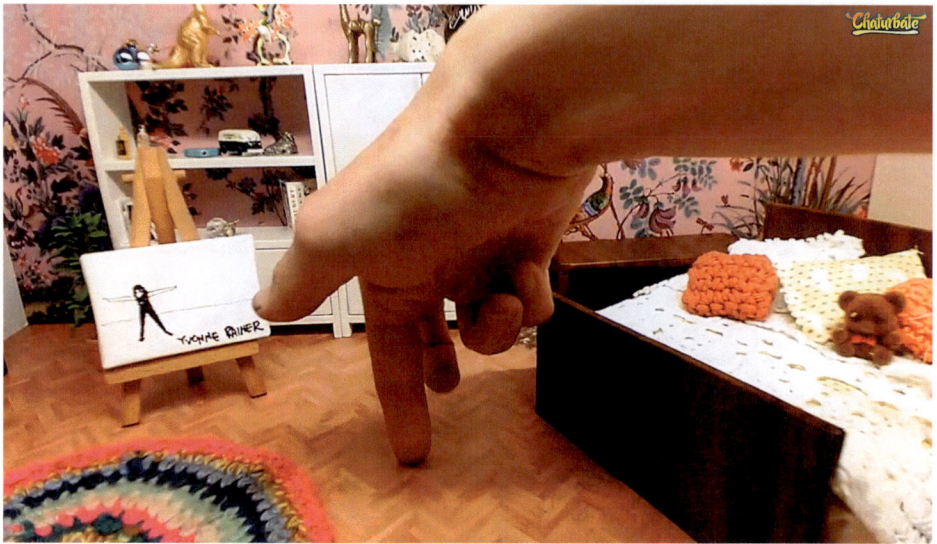

FIGURE 1.5. Screenshot of a live performance on Chaturbate, from *Maintenance Pornography* by Antonia Hernández, 2020

But I was trying to work on the platform (as a performer and researcher), not do something work-like. I had to find a way to combine my thoughts on maintenance and reproduction with a better awareness of what was happening on the sexcam platform without overimposing my ideas.

I was forgetting that the subversive aspect of Ukeles's maintenance art surpasses the visibilization of quotidian practices. Maintenance also discloses the fragility of the maintained thing, its tendency to decay and fail—and its creepiness. In this way, maintenance cannot stop, and it is only penultimately done. Even if the time of maintenance is always the present, it is a present that leaks into both the past and the future. The maintenance performed now preserves that something that was established before, being irredeemably conservative. But what is not maintained will perish, giving maintenance power over its futurity. What do the maintenance acts on the sexcam platform maintain, then? To what order are they loyal?

The temporal dimension of maintenance is not only linear. Incorporating the chance of accident, of decay and rotting, maintenance opens up the spectrum of the possible to many realities in which things explode and people have greasy fingers. Performing maintenance (and performing the performed one) is a way of being present to these many dimensions: what is here now, what happened, what could have happened, what it is intended to. In this way, while maintenance is conservative by nature, it also conjures disaster, and its absence is an open invitation to chaos. Maintenance, then, is about not only what was overlooked but what is impossible to capture. Any definite moment an illusion.

If the accident reveals the infrastructure, maintenance discloses its embodiment and relationality. As Susan L. Star affirms, "Infrastructure is a fundamentally relational concept, becoming real infrastructure in relation to organized practices."[10] Among those practices, maintenance encompasses the ones performed for the preservation of the infrastructure itself. This living, relational approach to structures discloses their mortality, showing them as temporary and precarious. Maintenance practices are hidden in plain sight not only because they are boring but because their visibility would reveal a secret: the work of the work, the fragility of the infrastructure, the decay of the platform.

Despite—or along with—my critique of the differential exploitation on the platform, there were more things happening there. People were sharing their time with strangers, discussing movie characters, exchanging recipes, having sex. Anna Tsing proposes to observe capitalism without the idea of progress.[11] What does it look like? she asks. It looks patchy. In this patchiness, precarity involves being vulnerable to others and leads to exchanges and generative contaminations. I realized that the idea of work was insufficient for encompassing all that was happening on the sexcam platform. I had, more or less, a set of tools for addressing the work of sex. I had to find a different strategy to speak about the sex of work: the togetherness, the excess, the aliveness.

I understood that I was also confused about who my audience was. I was performing for an imaginary future audience (artistic, academic) and not for the one I was facing in real time. If I intended to know something about working on the sexcam platform, even in that particular form, that was an unforgivable mistake. When I decided to perform for longer periods of time, improvising and interacting with the audience, things started to make sense. Two things happened then. The first was the evidencing of the audience as something that also requires maintenance. While a conversation starter with the public was generally related to the dollhouse, most people were just interested in having any kind of banal conversation, regardless of its setting. As I showed up daily on the platform, my work no longer was a reminder of invisible reproductive labor but involved many acts of maintenance—of myself, of the audience, of the platform.

The second major shift was the emergence of play. While I had been trying to play before, my play was highly constrained, limiting my actions to a safe

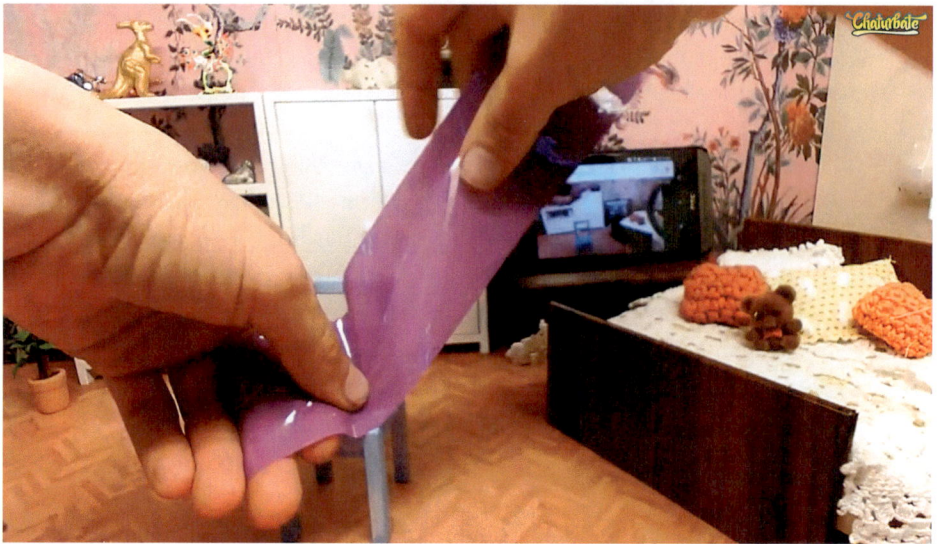

FIGURE 1.6. Screenshot of a live performance on Chaturbate, from *Maintenance Pornography* by Antonia Hernández, 2020

zone. As the loaded symbol that it is, the dollhouse was reigning and imposing a model. Dollhouses are complex entities, both places for the imagination and surveilled disciplinary apparatuses. It is exactly this conjunction that makes them recurrent objects for feminist performative revenge. It was, clearly, a paradox: I built the dollhouse, but I could not get out. A different play was possible, though. Its development revealed a richer ground with many unexpected encounters: among the participants, between the different genres and scales (1:1 and 1:12), and between critique and enjoyment—all in the same room.

Unsurprisingly due to the context, the play that appeared was mostly erotic, mixing miniature chair dances and (hand) bondage with housework. Addressing the rich territory of domestic porn, where things were interconnected and fluid, I was able to think in more performative actions while being truthful to the questions that animated the project in the first place. While I had a set of prerehearsed tricks, the audience became the most important factor: how they were reacting, whether they were laughing, whether I could keep them in my room. I understood that my insistence on building common ground was well intentioned yet misplaced. It was play—erotic or not—that allowed me to be vulnerable to others and open to encounters and contamination, to be present to that shared ground I was after. As Susanna Paasonen develops it, "My general

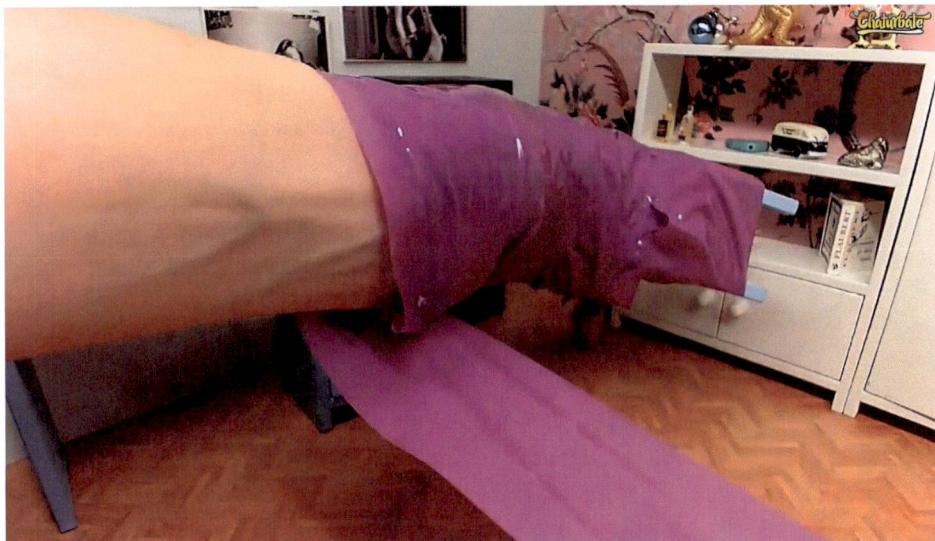

FIGURE 1.7. Screenshot of a live performance on Chaturbate, from *Maintenance Pornography* by Antonia Hernández, 2020

argument is that a focus on play makes it possible to highlight improvisation driven by curiosity, desire for variation and openness towards surprise as things that greatly matter both in sexual lives and in scholarly attentions towards them."[12] As a method and a disposition, playfulness allows the recognition of things that change—including bodies, habits, and structures.

This new approach also clarified what I was doing. Although I was using performance, this was better understood as a series of performative interventions. Well behaved but disruptive. By messing with the scale, by highlighting activities that are not supposed to be spectacular, I was denaturalizing the platform's habits and subverting them. This awareness lightened up my role and position as well. I did not have to pretend to be a performer—neither an erotic nor an artistic one. I was better equipped, however, for borrowing tactics from different disciplines and playing with them. I did not have to disentangle the overlap of sex, play, money, and reproduction but be able to be present to their messiness. This was one attempt at developing what I am calling a living method: not necessarily the right one and surely not the only possible one but one that can help in accounting for a broader range of experiences and fast-moving phenomena.

While these methods and the testimonies I gathered gave me important insights into the sexcam platform, I do not claim to have any experiences other than my own. I do not. However, the dollhouse allowed me to combine theory with a concrete and personal account of the platform and to go back and forth between them. In a sort of (small) feminist method, the dollhouse allowed me to acknowledge my position regarding the research, surfacing overlooked practices. Play also became a kind of queer method, able to embrace the awkwardness and instability of my position as both an insider and an outsider, privileged and precarious. I could sit comfortably with the creepiness of the platform.

Creepiness. That feeling that something is becoming too intimate, that something is forming a new arrangement in which you are, suddenly, participating. I propose maintenance, play, and maintenance play as activities to regain agency in that new compound. Through dedicated daily attention, maintenance does not show subjugation but reveals who is really in charge, how dependent things are on care, how fragile they are. Play allows engagement under agreement: revocable, changeable, pleasurable. And maintenance play creates a gap for maintenance, and play, to unfold.

NOTES

1. Keenan, *Technocreep.*
2. Here I am drawing upon Alys Eve Weinbaum's idea of the slave episteme. Weinbaum, *Afterlife of Reproductive Slavery.*
3. "Terms and Conditions," Chaturbate, accessed June 3, 2020, https://chaturbate.com/terms.
4. For more on the subject, see Viviana Zelizer and the idea of "hostile worlds." Zelizer, *Purchase of Intimacy.*
5. Ukeles, "Manifesto!."
6. Crary, *24/7.*
7. Haraway, *Companion Species Manifesto.*
8. Barthes, *Lover's Discourse.*
9. "No to spectacle no to virtuosity no to transformations and magic and make-believe no to the glamour and transcendency of the star image no to the heroic no to the anti-heroic no to trash imagery no to involvement of performer or spectator no to style no to camp no to seduction of spectator by the wiles of the performer no to eccentricity no to moving or being moved." In Wood, *Yvonne Rainer.*
10. Star, "Ethnography of Infrastructure."
11. Tsing, *Mushroom at the End of the World.*
12. Paasonen, *Many Splendored Things.*

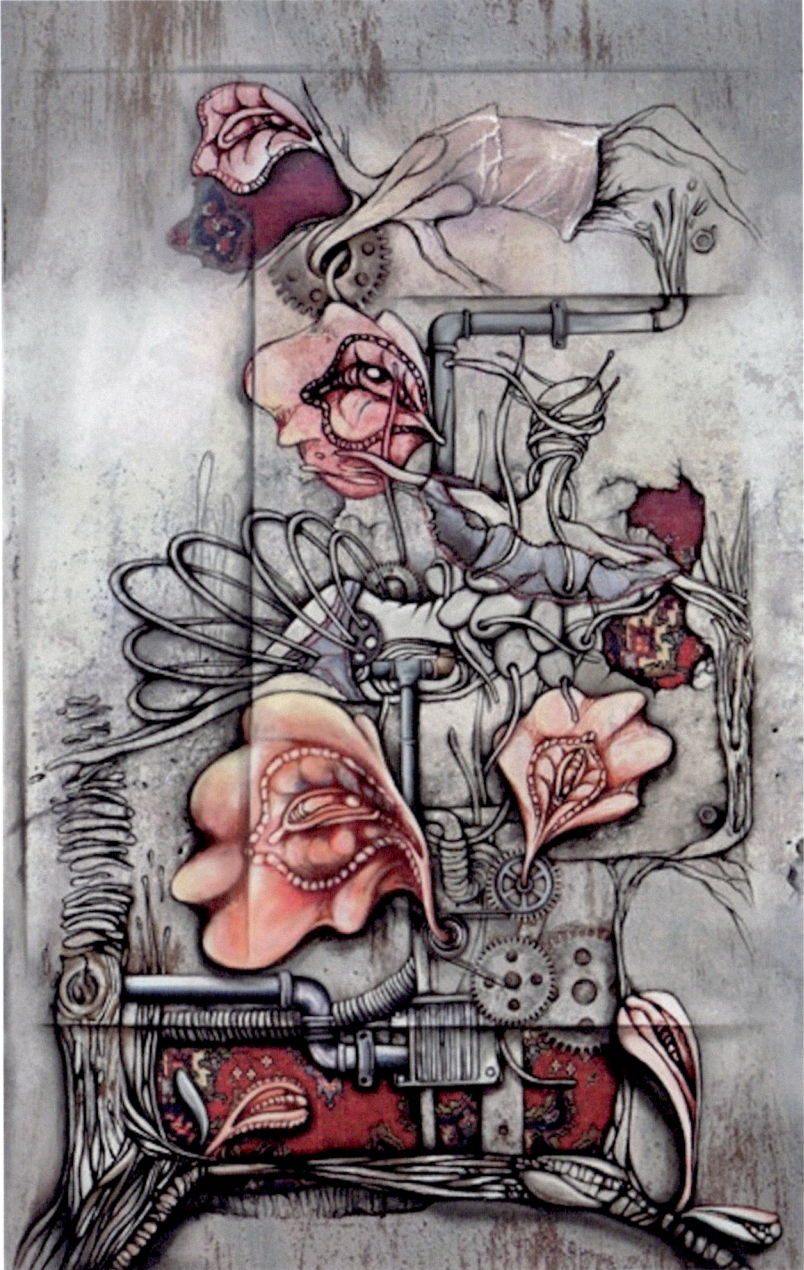

FIGURE INTER1.1. *The Embodied Self* by Marjan Khatibi

This work visually explores the complex relationship between technology and our embodied experiences. Using mixed media, the piece employs visual and compositional elements, especially texture, to explore the experience of engaging the whole body. The visual elements bind bodies to materials ranging from Persian carpets to industrial pipes, evoking a sense of touch. In doing so, the piece questions the ways that technology might inhibit our capacity to connect with our bodies and emotions. Representing the pain and hope inherent in the (feminine) body, the artwork illustrates how we might grow when approaching our experiences with curiosity. The multifaceted nature of the unseen senses embraces contrasting elements of discomfort and metamorphosis, encouraging a deeper understanding of resilience in the feminine self. The vibrant colors and genuine blossoms symbolize joy amidst hardship. The organic lines and repeated shapes signal growth versus progress. The piece captures the joy that resides in challenges through dynamic natural components and flowers. It explores the contrast between our physicality, embodied self, and the senses, revealing their interplay between technology, our connection with our bodies, and our innate growth through adversity.

NASSIM PARVIN

NEDA ATANASOSKI

Interlude

Smart Dust

Imagine a world where wireless devices are as small as a grain of salt. These miniaturized devices have sensors, cameras and communication mechanisms to transmit the data they collect back to a base in order to process. —John D. Sutter, "'Smart Dust' Aims to Monitor Everything"

Having all these things watching us could be super, super creepy and bad. And there's another layer to that, too. And how do you secure a trillion any-thing? How do you make sure that the nodes aren't lying, or infecting each other, or get hacked, or repurposed, or turned against us? —Bernard Marr, "Smart Dust Is Coming. Are You Ready?"

Dust is all around us, yet it rarely gets noticed. Dust is at once everywhere and nowhere, material and immaterial. It is imperceptible until it accumulates on a surface or gathers as a spectacular dust storm. These characteristics have made it a provocative vehicle for carrying the promises of smart technologies, and especially for the vision of ubiquitous computing idealized in a network of potentially infinite all-encompassing sensing nodes. In a 2010 piece on smart dust, CNN overviewed this new technology's driving vision to "monitor everything, acting like electronic nerve endings for the planet. Fitted with computing power, sensing equipment, wireless radios and long battery life, the smart dust would make observations and relay mountains of real-time data about people, cities and the natural environment."[1] Like its organic counterpart, smart dust promises to hold the capacity of stealthy and slow accumulation. Augmented with cameras, smart dust has the added capacity of seeing without being seen. And as such, it comes ready-made with both commonplace hopes and fears about the potential for granular monitoring. In a familiar manner of journalistic enthusiasm about technological novelty, an article in *Forbes* highlights many potential uses, such as monitoring crops, tracking inventory, and helping with medical and surgical procedures, alongside potential dangers, such as infringing on privacy and loss of human control. "Your imagination can run wild regarding the *negative privacy implications* when smart dust falls into the wrong hands. . . . The volume of smart dust that could be engaged by a rogue individual, company or government to do harm would make it challenging for the authorities to *control* if necessary."[2]

The tacit assumption in this technocratic imaginary is that smart dust is neutral. It can be put to good uses (surgical procedures, agriculture, etc.) or bad ones (a rogue individual that seeks to evade government control). This assumption elides the fact that all technology is inherently political. It makes nothing of its groundedness in governmental and political investments, needs, and values, or its military roots. In doing so, it falsely puts the burden of responsibility on actors who are conveniently categorized as good or evil. Questions remain, however: Who are the assumed ethical actors in this framing? And what happens if and when the neat separations that smart dust relies on break down?

Indeed, smart dust is neither neutral nor ahistorical. Smart dust depends on an ethos of "us versus them" for its conceptual and functional imaginary. Who and what needs to be seen, or vigilantly monitored, in the vision of smart dust? And who is trusted to see the data? In what new ways does smart dust bifurcate the planet? Or does it simply reaffirm the well-established categories and sedimented logics of empire that have crept into this seemingly new and

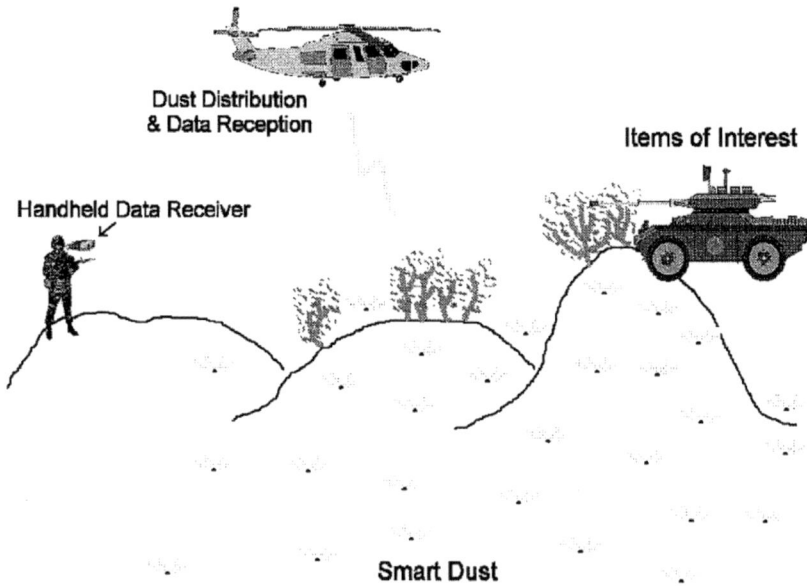

FIGURE INTER1.2. Rendering of smart dust in grant application

innovative technology? As a starting point, the temporal dimension of creep as accumulation is particularly useful in unpacking smart dust. It allows us to extend our thinking about the pasts and futures of this technology, as both matter and an ideal.

Tracing the history of smart dust we find ourselves, not surprisingly, in the world of accumulated military research and development. Its driving vision carries the fantasies of colonial expansion, occupation, and elimination of the enemy.

The original grant application that received funding from DARPA, authored by Kristofer Pister, envisions smart dust as an expansive network monitoring suspicious activity and dangerous weapons development by "rogue" nations. Furthering the police function of state power, dust occupies the most minuscule parts of enemy territory and reports on all activities. The grant application promises:

> With sensor motes containing acoustic, vibration, and magnetic field sensors it will be possible to cover many square miles of territory with a sensor network ... and report back a time history of all [movement] upon interrogation. . . . The sensor density required would be roughly 10,000 per square kilometer (ten meter average spacing), or in the range of thousands to tens of thousands of dollars per square kilometer. . . .

Once integrated into a Smart Dust system, [chem and bio] sensors will allow distant and early detection of chem and bio agents in combat. They could also be used to monitor precursor chemical usage at and around fabrication facilities in countries such as Iraq.[3]

Smart dust is a modern update to the eighteenth- and nineteenth-century British fantasy of an empire so vast where the sun never sets. Magnetic field, chemical, and biological sensors the size of dust particles promise a future where military and state actors can monitor and control every movement, down to the molecular level, in any territory in the world. Given the minuscule sizes of the many sets of electronic eyes, ears, and other receptors, this version of imperial control gets one step closer to total coverage while remaining invisible. Networks of motes can thus bolster various modes of occupation, even if that occupation is justified in the guise of humanitarianism.

It is no wonder that smart dust is identifiable as creepy, given that it is designed to be a known presence that evades observation. Yet, Michael Marder writes of dust as also a translucent medium of perception: "For Plato, the third element of sight interposed between the seeing and the seen was, itself, invisible such that its inconspicuousness could open a field of vision. Now, if this 'third' is populated with dust, then it is no longer completely imperceptible. Dust flakes hovering in a ray of light . . . reveal that ray . . . along with our lived space, never as transparent as we believe."[4] What, then, if we take smart dust, too, as a medium of perception? What does smart dust render visible in our spaces, our environments, and our ideals?

We might think, for instance, of the seemingly democratic and undiscriminating organic dust. Organic dust gathers on all surfaces. Yet organic dust does not remain or accumulate equally. Dusty spaces conjure abandonment and neglect. Clean and shiny surfaces require ongoing wiping and scrubbing. And it is through this activity that the discriminating side of dust surfaces. Dusting is the gendered and racialized work of maintenance: all too often invisible, underappreciated, unnoticed, and unpaid.

Smart dust is no different, as we may gather from another seemingly innocuous comment in Pister's grant. One of the largest obstacles to making smart dust a reality, he indicates, is figuring out how to "maintain the lifetime and maintenance cost of these investments."[5] In other words, and not unlike traditional modes of imperial command and control, smart dust will require constant maintenance and upkeep. Who will parse and clean the data? Who will pay the price for the messy data and the uneven accumulation of smart dust?

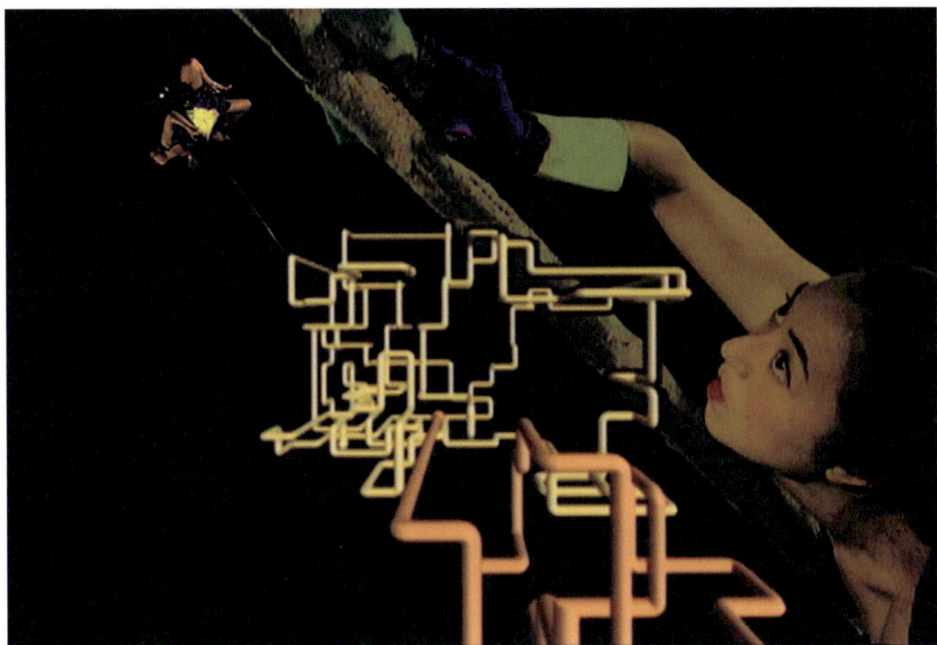

FIGURE INTER1.3. Screenshot, *Pipecleaner* by Katherine Behar

These questions conjure media performance artist Katherine Behar's 2007 project *Pipecleaner*, where she surfaces the often-invisible gendered and devalued work of technological maintenance. Behar's piece, a cross-platform screen saver, showcases pole dancers climbing, swinging, and gyrating on and across proliferating 3D pipes familiar to some viewers as the Microsoft screen saver that was ubiquitous in the 1990s. The dancers wear dresses made of cleaning gloves and wield sponges with which they scrub the ever-expanding network of pipes. Behar juxtaposes both kinds of gendered work as a feminist intervention. Pole dancers are meant to be gazed upon with desire. The work of cleaning is often put out of sight (for instance, after an office closes for the night or before guests arrive). At the same time, Behar puts on display the parallels in the repetitive and mundane nature of sex work and endless scrubbing. Both kinds of work "are constantly overwritten by the growing pipe maze."[6] Like dusting, the work of cleaning the infinitely but slowly expanding pipeline is never done. Countering the creep and accumulation of dust requires the ongoing attentiveness to those who perform paid or unpaid domestic work. Thus, in *Pipecleaner*, as the pipeline expands, so does the number of dancers on-screen.

The conflation of automated with gendered maintenance work on-screen fixes our gaze on the ostensibly dull, dirty, and repetitive work that has historically been devalued as unskilled, uncreative, and unworthy of our attention.

Yet, as Marjan Khatibi's art that accompanies this interlude conjures, there is both freedom and binding, pain and joy, that accompany maintenance work. Joy and resilience manifest in the vibrant flowers that spring from the industrial pipes. All of our senses, including touch, are at play in the distinct textures that come into collision yet coexist in the scene. In *Pipecleaner*, meanwhile, the juxtaposition of cleaning with dancing undermines assumptions about both while challenging our habits of seeing. Not only can we see the repetitiveness and dullness that characterize sex work, but we may also discern the erotic in cleaning and clearing, too. There is joy and fulfillment in maintenance, which moves away from capitalism's preoccupation with production. Not unlike cleaning, pole dancing is the ongoing work of maintenance: of bodies that perform, of interest in the performance, and of relationships with the audience. Maintenance is by no means unskilled, dispensable, or unimportant. Yet, we can note that in Pister's grant application, maintenance is framed as a minor if annoying aside to the innovation. The gendered and racialized notions of maintenance *creep* into the narratives of smart dust as an innovative technology.

Smart dust, like its organic counterpart, requires constant maintenance and (data) scrubbing. It too, when gathered, becomes waste. Because the smart dust units, or motes, are intended to be imperceptible, their conception and proposed design envisions single-use devices.[7] These tiny motes must each integrate sensors, circuits, communication transmitters, an antenna, and a power supply such as a solar cell or film battery. Smart dust would pollute the earth, air, water, and bodies in which it is dispersed. Unlike present-day electronic waste, which consists of large objects like cell phones or circuit boards, the waste that could be produced by smart dust may appear negligible at first. But herein is another way that we may see smart dust as creepy. Not unlike organic matter that creeps, we would only see the effect of smart dust as waste in time, when defunct motes gather. We may indeed choose to not see smart dust as e-waste at first, but it cannot evade perception forever.

Dust crosses boundaries. Organic dust crosses boundaries between the living and the dead. Smart dust crosses boundaries between intelligence and waste. And, most recently, smart dust is crossing the boundaries of organic and inorganic life. The newest research in smart dust technology resides in the fields of neural dust and body dust.[8] The goal of this new research is to realize "motes that could spread in [the] human body as thousands of individual sensors in a kind of sensing active network capable [of providing] telemetry from inside the body."[9] With the goal of personalizing medical care for the individual, body and neural dust would monitor everything from brain function to metabolic rates. In other words, smart dust comes with the hope of letting the body reveal its medical truths in unprecedented ways.

With smart dust, the Earth is also presumed to reveal its truths down to small particles. In 2009, the computer company Hewlett-Packard announced a project called Central Nervous System for the Earth (CeNSE). It planned to disperse a trillion sensors around the world.[10] With the goal of letting the earth *speak for itself*, researchers working on the project argued that "we're surrounded by technological assets that are deaf, blind, can't taste, can't smell and can't feel. . . . CeNSE is all about giving all this computer power the awareness of what's going on in the environment around it."[11] While the stated humanitarian intent was to monitor for healthy ecosystems, reduce energy use, ease traffic gridlock, and the like, the more immediate and commercial impetus for the project was for Hewlett-Packard to partner with the oil company Shell to build a wireless sensing and seismic imaging system. This would "allow Shell to more easily and cost-effectively explore difficult oil and gas reservoirs."[12]

In both of the above examples, smart dust could be seen as fulfilling the ultimate vision of ubicomp, where computing is made to appear. It's an idealization of an all-encompassing sensing and seeing apparatus offering the coveted *view from nowhere*, the fiction of a complete and objective perspective, at last! The sensing capacity of minuscule motes is matched by their power to linger. Dust cannot be eradicated—just redistributed.[13] The narratives of smart dust, however, continue to dwell on fantasies of gaining more efficiency, privileging oversight above embodied and localized ways of knowing, and the continued exploitation and use of natural resources.

Yet the question of what our bodies or our planet might tell us if they could speak to us opens vast possibilities for rethinking the relationship between the body, the mind, the spirit, and our environment. An imaginative reading of smart dust could steer us to the potential in the gatherings of always already partial perspectives. In each speck of dust, there is an occasion to see anew, and to embrace the partiality of sight. This does not need to be a limitation but an invitation to appreciate reflexivity and the value of collectivity and communion. After all, the intelligence of smart dust comes only from the collective and connected power of all the motes, or nodes. We might ask, then, What if we foreground the potential that might emerge from more expansive collaborations with smart dust? What other human-nonhuman alliances would be necessary or possible in the seemingly vacant spaces if we see their translucence with a beam of light?

NOTES

Epigraph 1. John D. Sutter, "'Smart Dust' Aims to Monitor Everything," CNN, May 3, 2010, www.cnn.com/2010/TECH/05/03/smart.dust.sensors/index.html.

Epigraph 2. Bernard Marr, "Smart Dust Is Coming. Are You Ready?," Forbes, September 16, 2018, www.forbes.com/sites/bernardmarr/2018/09/16/smart-dust-is-coming-are-you-ready.

1. Sutter, "'Smart Dust' Aims to Monitor Everything."

2. Marr, "Smart Dust Is Coming."

3. Kristofer Pister, "Smart Dust: BAA97–43," proposal abstract, University of California, Berkeley, accessed October 31, 2023, https://people.eecs.berkeley.edu/~pister/SmartDust/SmartDustBAA97-43-Abstract.pdf.

4. Marder, Dust, 20.

5. Marder, Dust, 4.

6. "Katherine Behar," Wikiwand, accessed October 31, 2023, www.wikiwand.com/en/Katherine_Behar.

7. "What Is Smart Dust and How Is It Used?," Nanowerk, accessed October 31, 2023, www.nanowerk.com/smartdust.php.

8. Carrara, "Body Dust."

9. Carrara, "Body Dust."

10. Sutter, "'Smart Dust' Aims to Monitor Everything."

11. Sutter, "'Smart Dust' Aims to Monitor Everything."

12. "Shell to Use CeNSE for Clearer Picture of Oil and Gas Reservoirs," Hewlett-Packard Corporation, accessed October 31, 2023, www.hpl.hp.com/news/2009/oct-dec/cense.html.

13. Marder, Dust, 4.

2

Uncivil Technoscience

Anti-immigration and Citizen Science in Boundary Making

It was one of those days, difficult to differentiate from the others that came before, when the sun exposed almost all of the entities of the Sonoran Desert to its violent force. Even in the shade, if they could find it, their bodies were still reached by UV lights and exposed to their negative effects. Under the cover of a saguaro or a creosote bush during the midday, partial shade could be lifesaving, though the superheated ground around them continued to be a reminder of the energetic reservoirs awaiting their flesh. To hide from the sun's sheer spectral power in this landscape was to embark on a losing fight. The vast, cloudless blue skies were a mirror image of the ways this landscape was bare to the all-seeing "eye" above. A whistling wind accompanied the whirring noise of air-conditioning units nearby when a robotic woman's voice announced: "Person Red." Near the Arizona-Sonora border, a seismic ground sensor installed by members of American Border Patrol (ABP) picked up vibration signals, which the barrier detection system broke down into data patterns. These data patterns led the system to identify the entity responsible for the vibrations: a human body traversing the desert. Geolocated data from the sensor were sent to a nearby small unmanned aircraft system (sUAS) waiting on a helipad. The barrier detection system automatically deployed the sUAS toward the source of the vibration signals, now named "Person Red." The color coding denoted the urgency of the sensor trigger and what this human figure represented to the system: imminent danger. The barrier detection system was patented by Glenn Spencer and Michael S. King for Border Technologies, Inc., a tech start-up founded in 2003.[1] Like the all-encompassing force of the sun casting its ultraviolet light over the Sonoran landscape and its entities, the barrier detection system made inconspicuous human movements perceptible by making certain kinds of data capturable even if their source hid in plain sight.[2]

Members of ABP set up their operational space in this spot in southeastern Arizona to research, develop, and test their various technological systems premised on a sense of care for the nation. Co-founded by Spencer in 2002, ABP is a registered nongovernmental organization that describes its mission as "monitor[ing] the border on [a] regular basis" and "educat[ing] the public about border control issues and solutions via internet and other media."[3] The Southern Poverty Law Center has designated ABP as a "hate group" for its anti-immigrant rhetoric, while Spencer has a long trajectory as an anti-immigrant activist, starting in the early 1990s when he created Voice of Citizens Together. In a 1999 video, this organization described how "an invasion is spreading across America like wildfire, bringing gangs, drugs, and an alien culture into the very heartland of America."[4] Voice of Citizens Together was involved in the fight to get Proposition 187 passed in California, which sought to institute a statewide citizenship verification system and prohibit unauthorized migrants from accessing nonemergency health care, public education, and other state services. Zeroing on ethnic Mexicans, Voice of Citizens Together and ABP are part of a larger network of xenophobic and racist organizations that imagine the US nation as under threat by a nonwhite, foreign, enemy force. The nation conjured by these organizations is a "heartland," a domestic space governed through racialized discursive practices of respectability that prioritize the inclusion of those (white) populations that can be assimilated and the exclusion of those deemed impossibly foreign.[5] Rhetorics of contagion and alien impurities have long been part of US anti-immigrant politics clamoring for the protection of the domestic sphere—both the home and the territory of the nation.[6] As American studies scholar Amy Kaplan shows, metaphors of the domestic are weighted tropes "imbued with racialized and gendered associations of home and family, outsiders and insiders, subjects and citizens."[7] Considered to be involved in "Civil Rights, Social Action, and Advocacy," ABP was classified by the National Taxonomy of Exempt Entities as a charitable organization. This chapter interrogates the purported charitability of this organization through the incivility of its technoscientific practices. To care for the nation, the group draws boundaries of belonging and expulsion and participates in the long history of nonstate actors enrolled to execute state violence.[8] The citizen at the heart of its organizational practices is a privileged subject position tending and protecting the purported purities of the imagined community—its peoples, culture, and territory.

I ask, What kind of citizenship is enacted in the technoscientific projects of nongovernmental organizations such as ABP? While its practices are often articulated in narratives of care for the nation, ABP reinforces the treatment of ethnic Mexicans, Latin Americans, and immigrants more broadly as "enemies" of the nation. Its care for the nation is premised on the technocreep of the white ethno-state, a nationalist populist project anchored in mistrust of government institutions, antiglobalist and anticosmopolitan visions, claims of defending working-class interests, and animosity toward immigrants and racialized others.[9] As Neda Atanasoski and Nassim Parvin argue in this book's introduction, one of the core dimensions of technocreep is its slow and gradual movement. The caring citizen and its technopolitical projects steadily ensconce and enact racialized, gendered, and settler colonial logics. White masculinist fervor is called upon to mark the spaces through might and wit where humanity is differentially produced and recognized. The fact that ABP and its operations were characterized as charitable—conducting "advocacy" and "civil rights" work—speaks to the methodical, imperceptible creep of such logics and their recognition and integration within the apparatus of the US empire-nation.

ABP contributes to the development of what I call the border technopolitical regime, a sociotechnical arrangement of entities enrolled in the production and governance of the material boundaries of the US nation. This chapter traces the collaborations between ABP, defense contractors, and the federal government in the articulation of a regime premised on a gradually growing, persistent technocreep—of data capture and processing, and bodily apprehension, displacement, and elimination. These collaborations take place through a citizen science framework that positions public expertise as outside of formal scientific institutions yet which contributes to the scientific enterprise. Doing so sheds light on the limitations of citizen science and calls into question its unanimous acceptance as a democratizing endeavor. I build on the work of scholars who have shown the relation between citizen science and neoliberal governmentality, and the capacity of civil society, generally seen as central to citizen science, to be uncivil.[10] I argue that, in the context of the United States, citizen technoscience has been central to the articulation of a white power sovereignty. The uncivil practices of ABP contribute to the border technopolitical regime's diverse modes of knowing anchored in sensing techniques predicated on removal—the extraction of data from unauthorized border crossers, their ejection from US territory, and their expulsion from the right to have rights.

Citizen (Techno)Science

Scholars in the social and natural sciences often treat the concept of "citizen science," first proposed by Alan Irwin, as falling into two camps.[11] Related to Irwin's original conceptualization of citizen science, the first camp highlights the need for scientific research to be democratized.[12] Irwin, concerned with questions of sustainable development and the environment in the mid-1990s, argued that public expertise was fundamental in the pursuit of scientific research related to environmental issues. Public expertise comprised the knowledge produced outside of the confines of professional scientific fields and their associated institutions. Citizens outside of the "formal" spaces of science worked with scientists who were engaged in tackling critical questions. Their interactions brought together vernacular and scientific knowledges that helped citizens respond to issues impacting their immediate communities. The inclusion of the wider public in a range of practices, like problem definition, data collection, and dissemination, not only expanded participation in the scientific enterprise but allowed scientists to examine previously ignored issues. Science moved from the enclosed space of the laboratory and into an area where it could be enacted by a range of previously excluded actors. The second camp of citizen science is one in which public involvement in scientific research is determined and managed by scientists. Caren Cooper and Bruce Lewenstein call this the contributory, or participatory, model.[13] Citizens might work on the collection, submission, or analysis of data, yet the methods, contours, and problems to be addressed all fall under the exclusive purview of scientific experts. The public is seen mostly as a source of volunteer labor (as in the case of amateur bird-watchers submitting information to the Cornell Lab of Ornithology) or, perhaps, as subjects of study themselves.

The democratizing model of citizen science has meant an increasing engagement between scientific and nonscientific groups around matters of concern. Commenting on David Hess's sociological assessment of the impact of globalization in the scientific enterprise, Sandra Harding contends that science now includes a range of practices that expand public engagement. In the social sciences, for example, researchers from within and outside the university collaborate through methods such as participatory action research to tackle scientific and technical problems affecting the latter's communities—particularly around issues of health and the environment.[14] Citizen advocacy is another realm within which members of the public raise new questions on existing science grounded in their experiences of various phenomena such as health. An emblematic example of this is the "lay expertise" of AIDS activists in the later 1980s and early 1990s, which helped reconfigure the boundaries of participation in scientific practice, the kinds of questions that could be asked and the kinds of research that could be funded, the treatment of participants, and more.[15] The contributions and importance of citizen science to contest and transform normative understandings of ways of being should not be undervalued. Still, other dynamics merit analysis.

Wider participation in science seems to suggest an expansion of who is included within the bounds of citizenship. To be a "citizen" who participates in scientific practice is to inhabit and enact specific relations—such as becoming a legible and validated subject by scientific practitioners—as well as to be produced by these relations.[16] Who has been included within the category of the citizen has varied across time and space. During the emergence of an independent United States in the eighteenth century, as American studies scholar Lisa Lowe argues, liberal colonial discourses steadily identified the rights of settlers, traders, and others to wage labor, free trade, and sovereignty, in contrast to racialized non-European peoples. These discourses described othered subjects "as unfit for liberty or incapable of civilization"; they were incapable of becoming citizens, of being subjects of the state.[17] And while, over time and through persistent struggles, incorporation within the boundaries of citizenship led to a more capacious figure, the "citizen" continues to be articulated in relation to an excluded other. This is what science studies scholars Neda Atanasoski and Kalindi Vora call the "surrogate human effect." The freedom of the modern citizen, the archetypal "liberal subject," "is possible only through the racial unfreedom of the surrogate."[18] The liberties awarded to the "citizen" are stitched to the unfreedoms of the Other. Citizen science is nested within this tethering.

Our present is marked by the unfolding of neoliberal capital, and practices of citizen science need to be understood in relation to it. In her book about food policing in post-Fukushima Japan, Aya Hirata Kimura shows how citizen science is part of a logic of accountability that puts "pressure on citizens to take responsibility for their own well-being and hence driv[e] them to collect data that are necessary for their health and safety."[19] The individual citizen is the starting point for neoliberal logics; all else follows this subject.

They are the agents of their own destiny, to paraphrase President George W. Bush, even if they are also the subjects of specific modes of governing that treats individuals and populations as living resources. Neoliberal political rationality is undergirded by moral and economic logics of individual responsibility and efficiency.[20] These logics help extend and disseminate market values to all kinds of institutions and spheres of action.[21] The need for self-reliance and self-sufficiency to supplement the social abandonment of a receding government apparatus has led to a growing network of nongovernmental organizations that pursue scientific projects to protect or support vulnerable communities and the environment. These organizations constitute the civil dimension of civil/citizen technoscience. In their work and as part of "civil society," they help further the machinations of the state by constituting the sphere through which consent is manufactured.[22] The fact that ABP pursues technoscientific projects on border enforcement and security, realms exceedingly well funded by the federal government, shows a different facet of citizen science in a neoliberal context. ABP's projects can sustain hegemonic conceptions of citizenship even as they further normalize and develop sociotechnical systems of governance.

There are those, like Luigi Ceccaroni, Anne Bowser, and Peter Brenton, who challenge the role of citizenship in citizen science. These authors claim that "the status of a person recognized under custom or law as being a member of a country . . . plays no role in 'citizen science.'"[23] As evidence, they cite the turn of various groups and organizations to rebrand citizen science under such terms as *community science, public science,* and *public participation in scientific research*. Yet this euphemistic redrawing of boundaries avoids addressing who is included and excluded in the making of science. Sidestepping citizenship in *citizen science* reproduces the purportedly universal project of liberalism that had, in fact, only successfully tethered the liberties of the citizen to the unfree conditions of the racialized Other.

Redeployments of Settler Techniques

The trope of a "flooded" border overwhelming US government officials is a recurring theme in debates about border and immigration enforcement, with particular intensity since the Cold War. Framing the border as "out of control," actors in government, the press, and industry justify renewed investments and commitments to treating some populations as "intruders," "threats," and "enemies" of the nation.[24] In the early 2000s, anti-immigrant paramilitary organizations like ABP pushed the US government to expand and innovate its security operations in order to counter the migrant tide. Redeploying the "Latino threat narrative," as anthropologist and border studies scholar Leo

Chavez suggests, these organizations denounced Latino migrants as not intending to integrate into dominant US culture. Anti-immigrant organizations imagined them as undoing US sovereignty by entering the nation's territory without authorization, violating the law.[25] ABP, the Minutemen Project, and other similar nongovernmental organizations used sUASs, streaming cameras, and crowdsourced labor from internet users to augment the reach of border enforcement.[26] These nongovernmental organizations should be understood not only as paramilitary organizations but as citizen technoscience ventures that emerge from the (neo)liberal program of the settler colonial state. These ventures position the citizen as responsible for democratically making the material implements and techniques that uphold the exceptions of imperial formations. Citizen technoscience enrolls the citizen as part of the border technopolitical regime and its reproduction of the state. The kind of citizen such a program makes possible is not just one that broadens participation but also one that is illiberal—one that works to police, discipline, and exclude the migrant Other. The "citizen" of technoscience, then, requires a critical disposition from the analyst to account for the creeping (as in gradual and slow-growth) boundary work such a figure is tasked to perform for the US empire-nation.

ABP was founded in 2002 in Arizona as the American Patrol, after Glenn Spencer left California, to fight back against what the group considered to be displacement caused by Mexican migrants. In a statement posted on the American Patrol's website, Spencer noted how "Americans, especially White Americans, should get out of California—now, before it is too late to salvage the equity they have in their homes and the value of their businesses."[27] Using the trope of an ensuing Mexican "invasion" bent on a plan for "*reconquista*" (reconquest), ABP imagined itself as protecting the US nation from this insurgency. In his video *detournements*, built from mixing snippets of interviews, congressional hearings, and news reports framed through overdub narrations, Spencer showed how this supposed *reconquista* unfolded, who was willingly or unwittingly supporting it, and who denounced it.[28] These videos can be read as part of the history of conspiracy theory narratives designed to get (white) "Americans" to rise up and fight back for "their" country.[29] In doing so, they fulfill expectations of the (neo)liberal citizen to be responsible and self-reliant, all the while maintaining their domestic sovereign space. The citizens in this situation are crucial, for it is they who are tasked with caring for and cultivating the US nation.

ABP's plan was to highlight cheap, effective means to enforce border security that the government could adopt. In a somewhat autobiographical and self-aggrandizing overview of the efforts of ABP and his role in the organization, Spencer said ABP was established to focus "on assessing how secure the border was, and efforts by the government to do the job. Particular attention was paid to technology."[30] Spencer's technological emphasis can be read in relation to his work experiences in operations research at a think tank in Washington, DC, and in geothermal energy and the oil industry. As the federal government discussed how to best tackle the new threat of post-9/11 international terrorism, conservative politicians and activists pushed for investments in border security and the recruitment of more agents for the reorganized Customs and Border Patrol (CBP) and the Transportation and Security Agency, both under the newly established US Department of Homeland Security (DHS). ABP began experimenting in May 2003 with attaching sensors (such as infrared) to remotely controlled fixed-wing aircraft. The group invited members of the press to the launch of its unmanned aircraft system (UAS) Border Hawk, which had cost them five thousand dollars to assemble and operate, and garnered local and some national media coverage. In doing so, the group convinced Arizona's congressional representatives to support the idea that UASs should be used in border security operations. Senator John McCain and Representative Jim Kolbe wrote to the secretary of DHS endorsing the idea.[31] Even though the sensors on the Border Hawk were not reliable and the drone itself was difficult to operate, by the end of the month, DHS announced "border drones" were indeed on the horizon of possibility for the agency.[32] In addition to flying over the rugged terrains of Afghanistan and Pakistan, military UASs would now creep into the domestic space of the nation, where they would direct their sensors on a range of border crossers, all treated the same—from unarmed migrants to coyotes and drug smugglers. The Border Hawk helped cement the reputation of ABP among a growing network of anti-immigrant paramilitary outfits in the Southwest. The citizen technoscience practice of ABP was a public relations success, because it got the federal government to implement a technological concept designed by the nongovernmental organization.

When it came to ABP's next project, Spencer relied on his experience in oil exploration to understand the usefulness of seismic sensors for monitoring human and nonhuman movements in the Sonoran Desert borderlands.[33] The use of these sensors, Spencer argued, was linked to a technical debate at the heart of border enforcement: how to measure the totality of unauthorized border flows. To answer this question was to propose "border metrics." Senator Elizabeth Dole told Congress in 2007 that "it's not just promises, it's proof that people want. The American people want to see results [in the US-Mexico border region], control of our borders. And we need to establish standards or metrics and then show that they've been achieved."[34] In citing the senator's declaration in his video on operational control, Spencer was legitimating the efforts of his group by showing how various political figures had been advocating for "border metrics," something ABP had been doing for many years. A Rand Corporation report from 2011 contended that "DHS leadership must define concrete and sensible objectives and measures of success." The Government Accountability Office agreed with Rand when it called on CBP to come up with "a performance goal, or goals, for border security between the [ports of entry] that defines how border security is to be measured, and a performance measure, or measures—linked to performance goal or goals—for assessing progress made in securing the border between [ports of entry] and informing resource identification and allocation efforts."[35] In other words, the Government Accountability Office's report makes an argument in favor of metrics as a means to measure the effectiveness of investments in border security. Border metrics become a feedback loop designed to govern the effective operations of border enforcement. By drawing on the words of elected and government officials, and technicians from think tanks, Spencer makes an argument for the need of objective measures that can be leveraged in the pursuit of operational control of the southern border, which, as the Secured Fence Act of 2006 states, means "the prevention of all unlawful entries into the United States." In doing so, Spencer insists on his barrier detection system as being strategically placed to offer CBP and government officials the metrics they have needed all along. And these metrics allow for clear goals to be set for the US Border Patrol and for its performance to be assessed in its pursuit and fulfillment of these goals. Cybernetic logics inform how border enforcement comes to be organized through the administration of information as technical matter.

ABP's intervention on the controversy of border metrics speaks to the growing synergy between this organization and the border technopolitical regime. Sensationalized coverage of these kinds of organizations occludes the gradual entanglements between anti-immigrant activists and government officials. In examining these entanglements through the framework of technocreep, we get to flesh out how the bodies of unauthorized border crossers become extractive ventures for defense industries and uncivil technoscience in the making of a technopolitical regime. In 2010, Spencer demonstrated a different technological concept to Arizona legislators. Back then he described his sonic barrier as more than just a detection system, because it could also be integral to a "border accounting system" that would allow actors to count the bodies that attempt entry into the United States without inspection on the border.[36] Human life in the desert borderlands was treated through the abstraction of numbers, of data to be captured, sorted, and fed back into the system for effective operations. The demonstration relied on members of ABP to perform the role of "unauthorized border crossers," their movements triggering the seismic sensors installed on Spencer's ranch. Legislators watched as data visualization from the sensors showed increasing noise, a sign that humans were approaching the barrier detection line. The elected officials were intrigued. This demonstration was part of Spencer and ABP's efforts to build relations with government officials, especially Republicans, and to show the benefits their technological systems could provide CBP in "securing" the nation. Spencer continued working on the sonic barrier system for Border Technology, Inc., with his colleague Michael S. King until they finally got a patent approved in 2015, with a second one granted in 2017. This second patent corresponds to the seismic detection and ranging mechanism (SEIDARM), which is connected to the sUAS called Hermes. SEIDARM comprises a network of continuous, solar-powered ground sensors collecting environmental data for potential human and vehicle movements in close proximity, such as low-flying aircraft. Each ground sensor has a range of about five hundred feet per one walking individual. The concept of operations for SEIDARM is based on an information system that records, processes, and mobilizes "border metrics" to guide border enforcement practices in "real time" and after the fact. The system's persistent technocreep functions as a "side arm," quickly accessible to the gunslinger in the necropolitical administration of bodies in the frontier.[37] Actors within this regime continue to draw from settler colonial fantasies of a wild frontier where unauthorized border crossers are imagined as indomitable dangers to the nation.

When Border Technology, Inc., presented the SEIDARM system at a US Air Force event in Las Vegas, Spencer found himself again showing how the citizen technoscience of ABP worked as part of the border technopolitical regime. The event in Las Vegas was part of AFWERX, a technology accelerator from the US Air Force meant to bring together innovators from within and outside the service. Spencer presented his system to AFWERX on April 6, 2018; engineers from Northrop Grumman were in the audience and offered Spencer the opportunity to integrate into SEIDARM a device developed by their company. It was the Mobile Application for UAS Identification (MAUI); installed on the Hermes UAS, this acoustic sensor would allow operators to identify other small drones nearby. And while SEIDARM/MAUI was tested for the folks at AFWERX in late June 2018, it is not clear whether the system was adopted by the US Air Force or is in use by another federal agency. DHS also invited Border Technology, Inc., to demonstrate SEIDARM/MAUI, but it appears nothing came out of the department's engagement.[38] Even though Border Technology's system is not part of the official arsenal of CBP, Spencer claims ABP has used SEIDARM to tell CBP agents where to intervene and apprehend unauthorized border crossers.[39] ABP's concept of operations, slowly developed since 2003, is strikingly similar to the Lattice platform and Ghost sUAS system developed by the Silicon Valley tech startup Anduril Industries and currently in experimental use by the CBP.[40]

At the heart of ABP's concept of operations we find a sociotechnical articulation of "vigilance" and citizenship. To be vigilant is to care for the nation. Thinking about this vigilant care through the work of feminist science and technology studies scholars on the politics of care means that to hold something in concern is to approach it in an attentive and thoughtful fashion. But this care does not necessarily entail a moral good. Instead, care lacks innocence, and much violence can be committed in its name.[41] Vigilance here is not an optical relation but an informational one. To sense, record, process, and communicate unauthorized border crossers as data—core processes I have elsewhere described as a cybernetic border—is and has been the dominant approach in US border enforcement operations since the end of the twentieth century.[42] What is significant in the uncivil technoscience of ABP is that it shows us how border enforcement, as a sovereign operation, is not limited to the actions of government officials. Border enforcement integrates a wide array of actors. And the citizen holds a privileged role for drawing the racial boundaries of the nation. The production of such boundaries unfolds through constant vigilance. This vigilant care is embedded within ABP's sensor networks and related to the performance of citizenship. Like the slow creep of "smart dust," information technologies are seemingly passive and harmless. But to be a citizen is to responsibly care for the nation by deploying information as the "difference that makes a *social* difference"; to be a citizen is to enact relations of informational dominance over surreptitious, racialized subjects.[43] To be a citizen is to exclude the migrant/alien Other. To subject them to the violence of an informational, multisensory regime designed to produce bodily capture, incarceration, and deportation. Uncivil technoscience democratizes boundary-making practices that make bodies and entities legible to the state and, consequently, excludable from its territory and the right to have rights.

These seeming (un)official interactions between CBP and ABP, as well as the multiple system demonstrations performed for state, federal agency, and military officials, demonstrate the uncivil character of ABP's technoscience. They seek to normalize and legitimize the treatment of unauthorized border crossers as "enemies" and "intruders" to be managed technologically. Renamed "Person Red," these crossers are sensed by a system that treats them as warning signs for a white ethnoracial order. Technoscience is recognized by the members of ABP like Spencer for its capacity to shape regulatory policies and modes of governance.[44] Science is taken as the final arbiter in the resolution of controversies over border security and border metrics. Considerations of sociocultural factors are seen as merely muddying the waters of the cool gaze of objectivity. In the eyes of ABP members, the historical relation between ethnic Mexicans and the Southwest or the imperial dynamics of global US racial capitalism are unnecessary considerations in the production of border security. Border security is, after all, associated with the preservation and reproduction of a kind of universal (white) sovereignty—the capacity of US society to institute a specific social and imaginary racial order. ABP's citizen science affords its members the capacity to perform the role of "constructive and helpful citizens—subjects who resourcefully take care of themselves" and of the nation.[45] They are paradoxically self-interested yet disinterested actors, objective in their care for the nation—represented by their conception of who belongs and who ought to be excluded from this imagined community.

Even when Border Technology is a separate entity, it is impossible to disentangle it from ABP. Documents from the Corporation Commission in Arizona show Glenn Spencer is registered as Border Technology's secretary and president. He is also the chairman of ABP's board of directors, where he joins former CBP chief agents (Bill King and Ron Sanders) in leading the organization. With elected state officials like former Maricopa County sheriff Joe Arpaio and state legislators attending many of the organization's technological concept demonstrations and social events, it is equally complicated to sever the ties that bind these communities together.

Conclusion

ABP is part of a contemporary uncivil society involved in the making of the nation-state. The group's incivility is made evident in its members' vitriol against ethnic Mexicans and Latinos more broadly. Construed as "threats," "intruders," and "enemies," these populations are impossible subjects beckoning constant vigilance. They justify and reproduce an enduring technocreep—a never-ending sociotechnical disposition that treats humans as the objects and subjects of data. The citizen technoscience of ABP is but an articulation of how the federal government enrolls and is enrolled by vigilante violence in the management of racialized populations. For some political scientists and government officials, ABP and Border Technology, Inc., are emblematic of what they see as the growth of nongovernmental actors involved in a changing landscape of drone operations and "threats" along the US-Mexico border.[46] But the creep of their operations is stowed away from these concerns. Technocreep can be found in the gradual growth and the persistence of government actor–nongovernment actor collaborations in border enforcement. The demo-cratizing narrative of citizen science finds its limit in these collaborations. They demonstrate that the liberties afforded to the citizen, both foil and product of technoliberalism, simultaneously making and being made by it, are of a partic-ular kind of racialized subject made legible to the state and another made un-recognizable. Citizen science in the context of US border enforcement opens up the category of "citizen" to critique and highlights its racial politics.

The harm inflicted on migrant communities branded as "enemies" of the nation are many. Chief among them is the way that sociotechnical systems like SEIDARM/MAUI are part of a vast assemblage that, since the development of "prevention through deterrence" in the 1990s, has been responsible for the death of thousands of border crossers every year. An American Civil Liberties Union report concluded in 2009 that "immigration policies have severely re-stricted legal entry, and border security policies have forced unauthorized entry through dangerous routes in perilous conditions."[47] This resulted in an esti-mated death toll ranging from 3,861 to 5,607 from 1994 to 2009. ABP's uncivil technoscience is not different from the systems developed and used by DHS in the production of harm—they are all part of the same "necroviolence" that instrumentalizes the desert environment.[48] Their practices work to normalize the pursuit of technological concepts against ethnic Mexicans and Latinos. As this chapter shows, ABP should be understood as the invisible infrastructure of the border technopolitical regime and its racializing politics of enmity.

And even as citizen technoscience is enrolled in the making of infrastruc-tures of enmity like ABP's SEIDARM, activists have also leveraged it to contest the failures of the state toward the human rights of unauthorized border cross-ers. In 2007, the Electronic Disturbance Theater 2.0/b.a.n.g. lab began devel-oping a mobile phone application to be used in cheap devices to guide people making their way into the United States through the deserts of the southern borderlands.[49] This "Transborder Immigrant Tool" used GPS technology to orient travelers in space by offering desert survival poetry and revealing the placement of water oases maintained by human rights activists. Working to counteract the necroviolence of DHS, the group of artists sought to enact a politics of radical hospitality—allowing travelers to overcome the lethal threat of dehydration and welcoming unexpected visitors seeking safety and refuge. Though the tool was never deployed and remains a conceptual technoscientific intervention, it highlights the ways that citizen technoscience is a multilayered practice in the making of specific kinds of citizenship. The task ahead of us is to describe the kinds of worlds citizen technoscience works to make possible and for whom.

NOTES

This chapter comes out of the work of the Border Tech Lab, first at Cornell University and now at the University of Texas at Austin. I'm grateful to BTL alumni—Alexandra Gutiérrez, Emma Li, Isaiah Murray, and Victoria Sanchez—for their efforts pursuing archival research for this project and their feedback on an early draft. I also want to thank Hanah Stiverson for her careful insights into the white power movement. Research for this project was supported by a Mellon Diversity Postdoctoral Fellowship in Latina/o Studies and Science and Technology Studies at Cornell.

1. Spencer and Davis, barrier detection system and method, US patent.

2. Glenn Spencer, "SEIDARM HERMES FOR NG," April 27, 2019, YouTube video, 3:49, www.youtube.com/watch?v=5kHcm1o8mMU. Many of the documents related to ABP were made by members of ABP itself and often tend to overvalue the group's success. My engagement with these materials is done primarily through attention to discourse and the articulation of imaginaries. Whenever I can't verify their claims through other independent sources, I state so.

3. NGO information obtained from "American Border Patrol," GuideStar, accessed October 31, 2023, www.guidestar.org/profile/42-1542666.

4. "American Border Patrol/American Patrol," Southern Poverty Law Center, accessed October 31, 2023, www.splcenter.org/fighting-hate/extremist-files/group/american-border-patrolamerican-patrol.

5. For more on respectable domesticity, race, and the US empire-nation, see Shah, *Contagious Divides*; and Wexler, *Tender Violence*. For migration, race, and US nation-making, see Ngai, *Impossible Subjects*; and Stern, "Buildings, Boundaries, and Blood."

6. Molina, *Fit to Be Citizens?*; Stern, *Eugenic Nation*.

7. Kaplan, *Anarchy of Empire in the Making of US Culture*, 3.

8. This practice stretches back into the nineteenth and early twentieth centuries with the Texas Rangers and the mob lynching of ethnic Mexicans in Texas in the 1910s. See B. Johnson, *Revolution in Texas*; Levario, "Home Guard"; and Muñoz Martinez, *Injustice Never Leaves You*.

9. Hugh Gusterson, cited in A. Chávez, "Gender, Ethno-nationalism, and the Anti-Mexicanist Trope," 5.

10. Kimura, *Radiation Brain Moms and Citizen Scientists*; Beittinger-Lee, *(Un)Civil Society and Political Change in Indonesia*; Chambers and Kopstein, "Bad Civil Society."

11. Ceccaroni, Bowser, and Brenton, "Civic Education and Citizen Science"; Cooper and Lewenstein, "Two Meanings of Citizen Science."

12. Irwin, *Citizen Science*.

13. Cooper and Lewenstein, "Two Meanings of Citizen Science."

14. Harding, *Objectivity and Diversity*, 16.

15. Epstein, *Impure Science*.

16. Chilvers and Kearnes, "Remaking Participation in Science and Democracy," 349–50.

17. Lowe cited in Atanasoski and Vora, *Surrogate Humanity*, 10.

18. Atanasoski and Vora, *Surrogate Humanity*, 5.

19. Kimura, *Radiation Brain Moms and Citizen Scientists*, 15.

20. Ong, *Neoliberalism as Exception*.

21. Brown, "Neo-liberalism and the End of Liberal Democracy."

22. Buttigieg, "Gramsci on Civil Society," 4–7.

23. Ceccaroni, Bowser, and Brenton, "Civic Education and Citizen Science," 9.

24. L. Chavez, *Latino Threat*; Chaar López, "Sensing Intruders"; Chaar López, "Alien Data."

25. L. Chavez, *Latino Threat*, 135–56; Navarro, *Immigration Crisis*.

26. Raley, "Border Hacks: Electronic Civil Disobedience and the Politics of Immigration."

27. "American Border Patrol/American Patrol," Southern Poverty Law Center.

28. Glenn Spencer, "Newt Gingrich Interviews Glenn Spencer 2001," July 14, 2019, YouTube video, 16:02, www.youtube.com/watch?v=uub98fQgOmY.

29. Belew, *Bring the War Home*; Stern, *Proud Boys and the White Ethno-state*; Uscinski and Parent, *American Conspiracy Theories*; J. Walker, *United States of Paranoia*.

30. Glenn Spencer, "My Border Report," American Border Patrol, 2019, accessed February 1, 2021, www.americanborderpatrol.com/IMAGES/190627/ATTACHMENT_1_MY_BORDER_REPORT.pdf.

31. Ignacio Ibarra, "Border Watch Takes to the Air," *Arizona Daily Star* (Tucson), May 1, 2003.

32. Spencer, "My Border Report"; Ignacio Ibarra, "Border Drones Possible by Year's End," *Arizona Daily Star*, May 23, 2003.

33. Spencer, "My Border Report."

34. Glenn Spencer, "NO 3 OPERATIONAL CONTROL," November 21, 2018, YouTube video, 6:52, www.youtube.com/watch?v=zd1EUU1Xr4M.

35. US Government Accountability Office, *Border Patrol*, 40.

36. Glenn Spencer, "Sonic Barrier Demonstration AZ Legislature," June 11, 2013, YouTube video, 15:00, www.youtube.com/watch?v=wcTUtrrMGXo.

37. De León, *Land of Open Graves*; Díaz-Barriga and Dorsey, *Fencing in Democracy*.

38. Spencer, "My Border Report."

39. Glenn Spencer, "SEIDARM AIDS BORDER PATROL," August 20, 2016, YouTube video, 3:49, www.youtube.com/watch?v=-z-_3-ru-Ro.

40. See contract no. 70B02C20D00000019 of July 2, 2020, between Anduril Industries and DHS, at a cost of $24.8 million for over forty sensor towers and at least one Ghost sUAS. Will Lennon, "The Virtual Wall: Documents Show CBP Plans for Surveillance Towers at US-Mexico Border," *Shadowproof*, April 8, 2021, https://shadowproof.com/2021/04/08/the-virtual-wall-documents-show-cbp-plans-for-surveillance-towers-at-us-mexico-border.

41. Martin, Myers, and Viseu, "Politics of Care in Technoscience," 627, 632.

42. See Chaar López, *Cybernetic Border*.

43. The idea of information as the "difference that makes a *social* difference" comes from Jonathan Beller's reformulation of Gregory Bateson. See Beller, *World Computer*, 13–14.

44. Sheila Jasanoff argues in *Design on Nature* that the normalization of science is central to neoliberal state-making.

45. Kimura, *Radiation Brain Moms and Citizen Scientists*, 17.

46. *Preventing Emerging Threats Act of 2018, Countering Malicious Drones: Hearing on S. 2836, Before the Comm. on Homeland Security and Governmental Affairs*, 115th Cong.

(2018) (statements of David J. Glawe, Undersecretary for Intelligence and Analysis, US Department of Homeland Security, and Hayley Chang, Deputy General Counsel, US Department of Homeland Security).

47. Jimenez, *Humanitarian Crisis*, 12.

48. Jason De León develops the concept of necroviolence to describe the operations of a human/nonhuman assemblage that performs "corporeal mistreatment" and instrumentalizes such mistreatment for "its generative capacity for violence." De León, *Land of Open Graves*, 69.

49. Electronic Disturbance Theater 2.0/b.a.n.g. lab. "Transborder Immigrant Tool." Net Art Anthology, https://anthology.rhizome.org/transborder-immigrant-tool, accessed November 1, 2023.

3

Hesitancy, Solidarity, and Whiteness

The Limits and Possibilities of Rape-Reporting Apps

In December 2016, journalist April Glaser published an article in *Wired* to alert the tech community about the gendered power dynamics that shape sexual violence reporting. On October 24, 2017, #MeToo began trending after actress Alyssa Milano used it as a Twitter hashtag in response to sexual assault allegations against Hollywood producer Harvey Weinstein—but it is important to note the phrase was created by African American women's rights activist Tarana Burke in 2006. Foregrounding a lesson of the viral #MeToo movement, Glaser emphasized the sexist assumptions informing how authorities disregard reports of assault and deny victims accountability: "Perpetrators can get away with attacking so many because often, when individuals report these crimes, they are ignored or discredited. But . . . after multiple survivors—previously unbeknownst to one another—shared the same story, the authorities and the public started to take their accusations seriously. There's strength (and credibility) in numbers, after all, but victims have few ways to connect with one another. We can fix this."[1] Glaser describes how an "ideal app" would support victims who want to come forward but whose motives are questioned. She argues such software could enable a victim to "communicate safely and anonymously with other victims—a mechanism for filing her story in a trusted, fully encrypted system that would allow her to maintain control of her identity." If other women reported the same perpetrator, the system could alert them to join a private messaging center to "communicate, coordinate, and, if they decide to, send a report to law enforcement and any organizations affiliated with their alleged assailant." Ultimately, as these women would come forward together, "police would be pressed to review the report, contact the victims, and weed out any impostors." Glaser concludes with a call to action, urging that while a version of this software exists on a few college campuses, women need a "more robust and universally available" solution that will materialize the unseen and unheard acts of sexual violence, making them more difficult for authorities to ignore.

While the extraordinary serial assault cases of the viral #MeToo movement foreground a core affordance that reporting apps offer—to connect multiple victim-survivors of the same perpetrator—it is the apps' evidence-documenting capabilities and promise of choice in when and how one discloses violence to authorities that serve the more frequent, everyday forms of sexual violence. Focusing on common sexual assault experiences, the ones without high-profile perpetrators and survivors and that are unlikely to capture media attention, better illuminates the relations that antiviolence technology materializes. Using technocreep as methodology to consider the complications and contradictions of technologically mediated reporting products, here I want to focus on questions that often go unexamined in public discussions: What values and ideas are inscribed in such apps' design? What forms of social and political power are taken for granted? And how do the racialized contours of sexual violence complicate the relations that technical reporting solutions promise? Examining these dynamics through the lens of technocreep can help unravel whom these apps are likely to serve and what inequitable outcomes they might reinscribe. In the Western settler-colonial context, cultural beliefs about sexual violence have existed in dynamic relationships with social hierarchies like race, gender, and sexuality. "Controlling images" predicated on the supremacy of white, heterosexual, and cisgender femininity have historically shaped knowledge of who counts as a survivor and what is legible as violence.[2] These narratives also shape dominant notions of accountability and justice *as* legal punishment, even though policing and prisons are a source of harm for survivors—especially for trans and queer survivors and survivors of color.[3] These politics of whiteness remain absent in conversations about reporting apps, raising concerns about how digital technologies will equitably empower survivors and foster transformative reporting justice.[4]

The Creep of Digital Technologies into Sexual Violence Prevention

The broader industry for digital antirape technology emerged under Barack Obama's presidency through a 2011 White House Office of Science and Technology Policy hackathon urging innovators to harness the power of mobile technology to prevent sexual violence.[5] Aneesh Chopra, then US chief technology officer, announced in a press statement, "We want to tap into the creativity of the American people to empower women who wish to communicate distress in a trusted and immediate way."[6] Heeding the call, designers have developed hundreds of antirape apps to provide users onetime emergency assistance, facilitate bystander intervention, and report and build evidence against perpetrators.[7]

The imagined role of digital technologies in sexual violence reporting is one rooted in addressing two well-established pain points of disclosing violence to institutional authorities. While affordances vary, reporting apps often match allegations against the same perpetrator in a database (e.g., Callisto, JDoe) or use artificial intelligence (AI) chatbots to interview victims and produce reports for HR to adjudicate (e.g., Talk to Spot, Botler.ai, #NotMe).[8] Proponents frequently laud such reporting apps for their alleged neutrality, endorsing them for their desired ability to produce powerful new types of evidence that will spark institutional and criminal legal accountability.[9] *PBS News Hour*, for example, praised Callisto for its ability to "meet the standard of evidence needed in sexual assault cases to make a conviction."[10] CNET, a popular technology forum, suggested that the JDoe app identifies serial perpetration "exponentially faster" than traditional reporting mechanisms in order to build strong cases for the civil law system.[11] And an article in *Fast Company* highlighted how Talk to Spot improves the "quality of evidence that complainants put forward."[12] While these endorsements underscore the fact that reporting apps are often in service of legal justice, the underlying need to remake disclosure conditions is profound. Understanding the politics of disclosure offers a way to contextualize why digital technologies have crept into the space of sexual violence reporting beyond a superficial account of rote technosolutionism.

Disclosing sexual violence is a critical moment in recovery and healing follow-ing an assault, yet it is notoriously unpleasant and overwhelming for survivors, who often feel a range of emotions, including fear, loneliness, anger, anxiety, depression, and preoccupation with the assault. Most survivors do not know what resources are available or how to engage them. When survivors receive insensitive treatment, they often hesitate to seek additional help and become further isolated.[13]

One mechanism contributing to survivors' hesitance is "credibility dis-counts," in which authorities systematically downgrade the trustworthiness and plausibility of violence accounts.[14] While survivors, in general, are subject to credibility discounts, the sociohistorical construction of victimhood that articulates *survivor* to middle-class cisgender feminine whiteness renders cer-tain survivors and perpetrators more legible than others. Put simply, survivors' ability to perform and embody idealized victimhood can become currency for institutional legibility. These durable politics of whiteness help explain a stark racialized inequality in reporting: for every white woman that reports her rape, five do not; for every Black woman that reports, at least fifteen do not.[15]

Whiteness is creepy in sexual violence disclosure—meaning it organizes social and legal relations in ways that mask how reporting decisions are not simple. Police rely on racialized myths and stereotypes about rape and extrale-gal factors, including victim age and race, to assess credibility, fueling unequal access to legal justice.[16] Legal scholar Sarah Deer recounts how because of legal loopholes on tribal land, "federal and state officials who had obligations to re-spond to crime in tribal communities simply ignored calls for help," leaving Indigenous women in a "vacuum of justice . . . as tribal nations have lacked the power or authority to prosecute crimes committed by non-Indians."[17] Be-yond legal gaps and apathy, due to long histories of targeted police violence, Black and Indigenous women and other women of color are more likely to distrust law enforcement.[18] Militarization at the US-Mexico border has also created conditions for border agents to assault immigrant women systemati-cally.[19] Investigations find police commit sexual assault at more than twice the general population's rate and target marginalized women, including lesbians, transgender and gender-nonconforming people, women and girls of color, women in the sex trade, and criminalized survivors who are more vulnerable to credibility discounts.[20] Hesitance around disclosure is thus not merely about individual survivor-authority dynamics but also about a broader system of gen-dered and racialized social control that shapes how authorities see and respond to violence.

The Creep of Whiteness in Reporting Apps

Given the creep of whiteness in reporting politics, what forms of techno-empowerment do reporting apps foster as they seek to enable new forms of survivor-authority relationships? Digital antirape technologies do not serve as passive survivor communication tools but actively shape knowledge about gender violence and how that knowledge flows across digital tools and institutional processes.[21] Put differently, culture and technology are co-constitutive, as scholar Anne Balsamo explains: "Through the practices of designing, cultural beliefs are materially reproduced, identities are established, and social relations are codified."[22] Callisto, JDoe, and Talk to Spot are three popular technologies used on college campuses, in corporate businesses, and by the broader public and comprise a valuable set for analyzing the values and forms of power mobilized through reporting apps. Callisto (projectcallisto.org) is the flagship software for college campuses that allows users to submit information about an assault to a "matching" algorithm to identify repeat offenders. JDoe (jdoe.io), launched in 2017, similarly uses an algorithm to identify repeat sexual offenders and enables users to "geocode" where assaults have taken place and to connect with a civil attorney. A third popular platform, Talk to Spot (talktospot.com), is an AI chatbot that conducts a virtual interview with survivors of workplace misconduct and harassment and sends the report to HR teams. Examining "how, for whom, and under what circumstances" reporting apps address disclosure can illuminate the apps' "creepiness" and the relationships between their technical features and offline effects on reporting politics.[23]

Techno-empowerment through Game Theory and Legal Action: Callisto and JDoe

Both Callisto's and JDoe's purpose is to use algorithms to technologically "empower" victims by addressing the problems of repeat offending, the underreporting of assault, and credibility discounts. Callisto's thesis is "If we can empower victims to take coordinated and informed action to protect their communities, we could prevent the majority of sexual violence in the United States."[24] JDoe similarly emphasizes that it "empowers survivors" to "act when [they're] ready, and not a moment before."[25] Both apps use information escrows to connect victims of the same perpetrator behind the scenes and eliminate what they identify as the "first-mover disadvantage," a concept from game theory that describes the high consequences faced by the first survivor to come forward.[26] The logic is that if someone is hesitant to be the only accuser, "then we should design a system that ensures that a person will not be the sole accuser and that his or her information will be aggregated with the information of others."[27]

Confronting the power dynamics that harm survivors is critical in addressing disclosure inequalities. Survivors too often are the object of institutional neglect rather than participating subjects in meaningful pursuits of justice. However, game theory, focusing on individual conflict in a zero-sum-game context, does not offer sufficient theoretical tools to account for how structural oppressions shape disclosure politics. Callisto aims to eliminate the so-called first-mover disadvantage, as those who use the app's "matching option" only have their cases sent to authorities when there is a second allegation. However, credibility discounts are not rooted in the number of survivors to come forward. As neither app communicates, contextualizes, or designs for how people's social identities shape access to justice, this becomes one crucial way reporting apps articulate political whiteness, meaning an "orientation to and mode of politics that invokes state and institutional power to redress personal injury, while erasing how sexual violence is experienced in a white-patriarchal apparatus."[28] Moreover, as critical observers point out, many people *"can't* call the police because of . . . fear of deportation, harassment, state-sanctioned violence, sexual violence, previous convictions or inaccessibility."[29]

Despite these observations, Callisto offers a normative vision of disclosure whereby providing multiple accounts of serial perpetration will prompt authorities to act quickly and equitably. Marketing narratives emphasize the company's commitment to empowering survivors by providing new technical "options."[30] While Callisto links to information on nonlegal resources like self-care and mindfulness apps, its matching algorithm is the real product. The matching algorithm allows users to submit information about an offender's identity and store it in Callisto's encrypted database. When two or more users name the same offender, a match occurs. Callisto provides users access to their matching algorithm "under the precondition that, if a match to another case by the same perpetrator is found, a Callisto Legal Options Counselor™—an attorney contracted by Callisto—will reach out to each victim individually."[31] Here, the kind of "empowerment" offered is further filtered through the lens of legal justice and the tacit assumption of political whiteness that legal systems effectively work for all victims.

JDoe's invitation to choice and collective action similarly encourages users to pursue recourse through the civil law system. JDoe's objective to "link victims of mutual offenders with lawyers" is meant to streamline the process for reporting misconduct and assist personal injury firms in finding compelling class-action cases. In an interview with the *New York Post*, Ryan Soscia, who developed JDoe in 2015 after discovering that the same sports trainer had assaulted him and his teammates, explains, "There's power in realizing you're not alone . . . and that could be a powerful throughput for the justice system."[32] While the company suggests that "cases with multiple plaintiffs are less harrowing for victims and more likely to win in court"—a popular truism—it similarly relies on the presumption that the civil law system works equitably if victims are simply better organized. This belies the fact that progressing a class-action lawsuit poses hurdles. There must still be a representative plaintiff, for example. While the specific rules governing class-action lawsuits vary by state, the civil court judge still holds significant discretion in certifying the class so that the case can proceed. Certification includes the judge finding that the representative plaintiff has suffered the same harm as other members of the proposed class, that there are enough members of the proposed class (e.g., often at least twenty-one victims, in the United States), and that a class-action lawsuit is the proper way to resolve the claim.

Rather than mobilize more transformative relations of solidarity and support outside legal systems, the JDoe app is optimized to rapidly collect basic information needed to construct a class-action lawsuit. Users document their experiences via drop-down buttons to select "rape or attempted rape," "unwanted touching," or "sexual harassment" and specify their relationship to the perpetrator—whether he, she, or they were a partner, family member or guardian, colleague or employer, teacher or coach, acquaintance, or stranger. Users provide identifying information about the perpetrator, such as a phone number or a link to social media information, and are given a choice whether or not to send an anonymized version of the report to a participating law firm. However, this "choice" is not much of a choice at all, as the app's primary benefit is connecting users to civil lawyers who pay for access to victims' names and information. While JDoe asserts that its product "circumvents authorities who may not have victims' best interests in mind," JDoe effectively sells victims' data to civil law firms—and takes a cut of any settlements.[33] Given JDoe's (and other reporting apps') entanglements with venture capital funding, claims about upholding victims' best interests might justifiably come into question. And without raising the question of, or accounting for, the cultural logics that contribute to how civil law authorities similarly privilege whiteness, JDoe encourages an unexamined endorsement of civil law recourse.

Identity-Neutral Chatbots and Gold-Standard Evidence: Talk to Spot

Spot aims to "create more inclusive workplace cultures" by "making it safer for employees to speak up and easier for HR teams to take action."[34] Spot's AI chatbot interviews employees about incidents of harassment in the workplace, and according to its cofounder psychologist Julia Shaw, PhD, Spot is designed to follow the cognitive interview technique that police use to interview crime victims. The company's decision to affiliate the AI with law enforcement communicates its affinity for legal practices of evidence production—rather than for the approaches of victim advocates or psychologists whose work was formative in uncovering how trauma shapes memory.

However, what is most concerning is the company's dubious claims that its AI is a completely "unbiased bot" and that "because Spot is a bot and not a human, it cannot judge or assess you."[35] In framing the AI as a "neutral mediator," Spot fails to acknowledge that chatbots are never unbiased—as they are designed by people and trained on data that encode ideas and values into the system.[36] The ways in which AIs exhibit white voice, language, and cultural sensibilities are well documented.[37] As scholar Zoe Vorsino notes of chatbots in general, "Any simulation of human intelligence via communicative software is also inherently a question of race, class, and gender," as users communicate with chatbots through their general assumptions about the world.[38] Rather than recognize these realities, Spot blackboxes its development processes, failing to explain how the company trains its AI not to reproduce cultural beliefs contra an inclusive reporting experience, such as the endorsement of racialized rape myths.

Spot does, however, emphasize that its AI is attuned to how trauma shapes communication and "how easy it is [for victims] to not remember important details."[39] From this understanding, Shaw determined that "people needed a way to contemporaneously create a really high-quality piece of evidence that they could use if they wanted to take it further and make a complaint to their organization." Shaw wants to redefine Spot as the HR "gold standard" in reporting violence, claiming that a record from Spot would be as "high quality as you can get, basically."[40] Yet while the AI facilitates disclosure, the actual outcomes remain contingent on the humans who act on the information the AI collects. The company never explains how it challenges the likelihood that human reviewers of these reports may filter them through rape myths or subject employees to credibility discounts—even when a chatbot has conducted the interview. Without attention to these issues, Spot's AI does not reform institutional behavior; instead, it disciplines survivors to render reports of violence more navigable to authorities.

Designing for Power and the Creep of Whiteness

Examining reporting apps through the lens and method of technocreep reveals their complications and contradictions. On the surface, reporting apps reflect the ideals of critical design scholarship. Certainly, reporting apps attempt to address the power dynamics of reporting and spin out new ways of relating to institutions. Much antiviolence tech is developed by survivors and advocates who are outsiders to institutions. Their efforts to empower seek to increase agency and give survivors collateral to navigate power-laden institutions. They work to advance the right to bodily integrity—the idea that people have the right to be free from gender violence.

However, while proponents maintain that these technologies will correct situational sites of injustice, reporting apps still reflect what Ruha Benjamin calls the "duplicity" of technology—the idea that technology can in some ways serve justice but often adopts the default settings of our society.[41] While reporting apps aim to "empower" through new, technically mediated relations, they prioritize normative reporting methods and confront most directly the institutional frictions primarily faced by white, middle-class victims of harassment and violence. In doing so, they reaffirm and legitimate the logic of political whiteness that victims should report their assaults for the primary purpose of engaging formal institutions: by using AI to create the "best account" for HR managers (Spot), by using information escrows to enable victims to come forward together for legal action (Callisto and JDoe), and by allowing victims to delay the submission of their reports until they (ultimately) decide to come forward to institutional authorities (Callisto, JDoe, and Spot).

While these apps may relatively improve reporting experiences regardless of one's social location, the technorelations they engender align to dominant power dynamics. In other words, the apps' political reformations are one-sided. They fail to substantially change how institutions interact with victims—changing instead how victims interact with institutions—and thus fail to fully understand institutions as structures of power. By operationalizing disclosure politics in the neutral and unmarked terms of economic game theory, the platforms become an implicit space of whiteness. None of the apps advertise the outcomes and benefits of using their software. In digging deeper, we can begin to see how instead of radically reimagining justice, the reporting apps pitched as power-shifting technologies are more likely to serve corporate and legal gazes by disciplining disclosure according to the imagined needs of white, middle-class, straight victims and white, racialized institutions. In short, while they correct some injustices, they create new justice issues.

So how do we get closer to that promise of technology and create interventions that embody, materialize, and reimagine justice practices? Engaging technocreep as an analytic can help uncover and anticipate the felt and unseen social power dynamics that constitutively shape a technology's impact in the world. Designing for power requires confronting the multifaceted creep of whiteness through radical feminist and antiracist ways of seeing. One method is to think of our values not as inputs that fit neatly into a technological equation but as experimental hypotheses for rebuilding the kind of liberatory world we want to live in.[42] Designers must approach reporting apps and other antiviolence technologies not as neutral tools but as experiments that can give form to or reconfigure the legacies of patriarchal violence, white supremacy, transphobia, and colonialism. Efforts to design justice must not focus on disciplining victims' behavior but on critically thinking about the power structures that shape disclosure. It is only through such a politics of resistance and visibility that antirape technologies can challenge the creep of political whiteness.

NOTES

1. April Glaser, "How Simple Software Could Help Prevent Sexual Assault," *Wired*, December 9, 2016, www.wired.com/2016/12/prevent-sexual-assault-with-software.

2. Collins, *Black Feminist Thought*, 69.

3. Ritchie, *Invisible No More*.

4. Shelby, Harb, and Henne, "Whiteness in and through Data Protection."

5. Beaton, "Safety as Net Work," 105.

6. Kathleen Sebelius, "'Apps Against Abuse' Challenge to Help Address Sexual Assault and Dating Violence," White House blog, July 13, 2011, https://obamawhitehouse.archives.gov/blog/2011/07/13/apps-against-abuse-challenge-help-address-sexual-assault-and-dating-violence.

7. Eisenhut et al., "Mobile Applications Addressing Violence against Women."

8. Henne, Shelby, and Harb, "Datafication of #MeToo."

9. Sim, "Respond and Resolve," 8.

10. Jess Ladd, "How I Help Sexual Assault Survivors Feel like They're Not Alone," *PBS News Hour*, aired October 19, 2017, www.pbs.org/newshour/brief/231133/jessica-ladd.

11. Jonathan Grieg, "Reporting Sexual Misconduct Could Get Easier with New Apps JDoe and Callisto," *CNET*, August 22, 2018, https://download.cnet.com/news/reporting-sexual-misconduct-could-get-easier-with-new-apps-jdoe-and-callisto.

12. Anisa Purbasari Horton, "These Apps Try to Make Reporting Sexual Harassment Less of a Nightmare. Do They Work?," *Fast Company*, February 12, 2019, www.fastcompany.com/90303329/these-apps-try-to-make-reporting-sexual-harassment-less-of-a-nightmare-do-they-work.

13. Jacques-Tiura et al., "Disclosure of Sexual Assault."

14. Tuerkheimer, "Incredible Women," 1.

15. Slatton and Richard, "Black Women's Experiences of Sexual Assault and Disclosure," 2.

16. Venema, Lorenz, and Sweda, "Unfounded, Cleared, or Cleared by Exceptional Means."

17. Deer, *Beginning and End of Rape*, 41.

18. Ritchie, *Invisible No More*.

19. Falcon, "Rape as a Weapon of War," 33.

20. National Police Misconduct Reporting Project, *2010 Annual Report*, 3; Purvis and Blanco, "Police Sexual Violence," 1496.

21. Bivens and Hasinoff, "Rape," 1064.

22. Balsamo, *Designing Culture*, 3.

23. Quote in J. Davis, *How Artifacts Afford*, 11.

24. Anjana Rajan, "Callisto: A Cryptographic Approach to #MeToo," presentation given at Enigma: A USENIX Conference, 2019, YouTube video, 16:58, https://youtu.be /naLTSqGQ_zY. Quote at 2:45.

25. "Empowering Survivors," JDoe, accessed September 30, 2021, https://jdoe.io/html /features.html.

26. Rajan, "Callisto," 2:25.

27. Michael Chwe, "How Economic Theory Can Help Stop Sexual Assault," PBS, December 19, 2014, www.pbs.org/newshour/nation/economic-theory-can-help-stop -sexual-assault.

28. Phipps, "'Every Woman Knows a Weinstein,'" 1.

29. Mia Mingus, "Transformative Justice: A Brief Description," *Transform Harm*, January 11, 2019, https://transformharm.org/tj_resource/transformative-justice-a-brief -description.

30. "Explore Your Options," Callisto, accessed September 30, 2021, https://www .projectcallisto.org/should-i-report.

31. "Terms of Service," Callisto, accessed September 30, 2021, https://www .projectcallisto.org/terms-of-service.

32. Marisa Dellatto, "People Are Using a New App to Report Sex Assault Anonymously," *New York Post*, September 21, 2018, https://nypost.com/2018/09/21/people-are -using-a-new-app-to-report-sex-assault-anonymously.

33. Quote in "Are You a JDoe?," JDoe, accessed April 10, 2020, https://jdoe.io/html /about.html.

34. "The Research behind Spot," Talk to Spot, accessed April 10, 2020, https:// talktospot.com/research.

35. Sam Mercer, "Spot, Report, Stop: AI Tackles Age-Old Problem," *Counsel*, October 21, 2019, www.counselmagazine.co.uk/articles/spot-report-stop-ai-tackles-age-old -problem.

36. For "neutral mediator," see Stefanie Dorfer and Elspeth Taylor, "AI Takes Bias Out of Workplace Harassment Reporting," Stylus, October 23, 2018, www.stylus.com/ai-takes -bias-out-of-workplace-harassment-reporting.

37. Phan, "Amazon Echo and the Aesthetics of Whiteness."

38. Vorsino, "Chatbots, Gender, and Race on Web 2.0 Platforms," 113; Marino, "Racial Formation of Chatbots," 5.

39. Emily Reynolds, "How Technology Is Tackling the Stigma around Sexual Assault," *Vice*, March 14, 2019, https://web.archive.org/web/20220818172732/https://i-d.vice.com /en_uk/article/mbzw9a/technology-sexual-assault-apps-spot-callisto.

40. Reynolds, "How Technology Is Tackling the Stigma."

41. Benjamin, *Race after Technology*, 154.

42. Jafari Naimi, Nathan, and Hargraves. "Values as Hypotheses."

4

Undoing Landlord Technologies
Beyond the Propertied Logics of the Pandemic Past and Present

In mid-2020, amid ongoing housing struggles exacerbated by the COVID-19 pandemic, I met up with a close friend in San Francisco's Mission District for what had become one of many normalized six-feet-apart walks. As we slowly meandered up nearby Bernal Hill to look down on the quarantined city, my friend began telling me of their latest trials and tribulations with housing. Without explanation or consent, their landlord had installed a Google Nest facial recognition camera complete with audio capture capacity right outside of their bedroom window. It is now fastened to the building's infrastructure, peering down onto the front porch below while picking up sound through my friend's thin, rattly West Coast window.

As I soon came to find out, amid the pandemic present and the ongoing uprising for Black lives, Dish TV had begun selling integrated Google Nest cameras to property owners under the racist auspices of protecting buildings from "dangerous looters" allegedly roaming the streets. While there had never been any property damage or theft inflicted on my friend's landlord's building (except apparently for one newspaper stolen from the front porch back in the 1990s), he, like many property owners, bought into white supremacist portrayals of threat stoked in the wake of George Floyd's death. These imaginaries have been used to entrench carceral technologies into racial landscapes, with Google Nest—like Amazon Ring, Flock Safety, and other camera-based home and neighborhood surveillance systems—sold as a remedy to anticipatory property damage. Such companies have been known to hand footage over to law enforcement agencies, which then, often enough, interpret and act upon biased data.[1] This transpires amid the rise of neighborhood "snitching" apps such as Nextdoor, Citizen, and even 311, in which homeowners and neighbors deputize white, wealthy, and imperial visions of threat through digital technologies.[2]

While racist surveillance landscapes were alive and well prior to both the pandemic and the uprising for Black lives, new forms of domestic surveillance became legitimized in their wake under the auspices of suppressing danger, along with pandemic-related contact tracing, home employment and schooling platforms, and virtual property management. New landlord technologies were introduced into an already racist housing landscape in order to curb the advance of threat/virus. In this way, landlord tech today anthropomorphically combines familiar playbooks of disaster capitalism with racial capitalist technologies to maintain security. To understand this process, it is helpful to think with Ruth Wilson Gilmore's observation that prisons and the broader prison industrial complex (which includes homes and neighborhoods) function to defang the potential threat of surplus populations rendered idle and interpreted as dangerous to capitalist production—a context that the pandemic, alongside racial justice organizing, imbued. Carceral technologies aim to reproduce capitalist stability, especially amid crises, empowering societies to "lock people out by locking them in."[3]

By mobilizing these carceral logics, landlord tech not only is used to keep imagined threat from entering property's domain, but additionally it locks people out from the ability to rent housing through tenant screening systems that blacklist those with prior eviction, criminal, or poor credit records from securing future housing. As a form of landlord technology, tenant screening celebrates the imaginary of the "good" tenant while discarding the possibility of stable housing to those already disavowed through racial capitalist calculations. In pandemic contexts, landlord tech also serves to bar the racialized figure of what Chikako Takeshita describes as the "virushuman" from contaminating the fantasized purity of private property.[4] In this sense, the pandemic conjuncture gets compressed into what Stuart Hall once described as a policeable crisis,[5] in which racialized imaginaries of monstrous threat can be curbed by employing invasive technologies. Such a crisis transpires when social formations can no longer reproduce through pre-existing social relations, and when the dispossessed, suddenly a threat to capitalist order, are compacted into the mutable figure of the viral monster.

Contemporary landlord technology employs algorithms, platforms, and artificial intelligence (AI) to keep this monster at bay, yet it also galvanizes a deeper history of private property itself functioning as a racial technology of surveillance, differentiation, and dispossession—whether applied to land, bodies, or data. This fuels the engine of what I describe as racial technocapitalism, an analytic that homes in on how technologies of racialization and carcerality abet the work of capitalist reproduction by suppressing the threat of rebellion. As racial technocapitalism functions through imperial and white supremacist relationships to property—encompassing everything from the parceling of colonized lands to the instantiation of racial surveillance technologies, data regimes, and infrastructures—it is an apt conceptual lens through which to theorize landlord technologies.

As a dispossessive logic, racial technocapitalism takes into account how the reproduction of private property regimes has long required technologies of categorization to delineate who and what can be owned or incarcerated, and by whom. It builds upon what Brenna Bhandar describes as "racial regimes of ownership," or "economic visions of land and life rooted in the logics of abstraction, culturally inscribed notions of white European superiority, and philosophical concepts of the proper person who possessed the capacity to appropriate."[6] Such regimes require scientific technologies of quantification, measurement, and analysis to demarcate and uphold the logics of what Cedric Robinson defines as racialism, or a material and cultural dynamic (or perhaps technology) intrinsic to capital's need to legitimize the hierarchical property relationships that have long haunted Europe and its colonial geographies.[7] As Neda Atanasoski and Kalindi Vora theorize, race then functions as "the condition of possibility for the emergence of technology as an epistemological, political, and economic category within Euro-American modernity."[8] Technology, in other words, works as a tool of racial differentiation and dispossession, while racial technocapitalism helps articulate its uneven manifestations, including in the space of home.

In this chapter I mobilize the analytic of racial technocapitalism to map landlord technolgical creep into domestic space amid the capitalist crises of the pandemic past and present. Looking to the industry's growth throughout the Cold War and its subsequent expansions during the dot-com boom, 9/11, the subprime mortgage crisis, the AI boom, and the COVID-19 era, I offer land-lord technology and its dispossessions a contemporary genealogy. Yet at the same time, I chart possible ways out, looking toward abolitionist, feminist, and decolonial approaches to what Ananya Roy calls "undoing property."[9] I also investigate strategic countermappings of landlord property portfolios and land-lord tech platforms themselves, focusing on collaborative work I have been a part of with the Anti-Eviction Mapping Project to reposition property data into the hands of tenants and tenant organizers outside of and despite racial technocapitalist property logics.

Landlord Tech

Contemporary landlord technology rests upon a thick palimpsest of private property regimes and is often paternalistically implemented under the supposition of caring for tenants and neighborhoods. Yet often enough it privileges care for buildings and their value more than for those residing within them. Its surveillance platforms meanwhile extend the carceral state into domestic space, getting at what abolitionist scholars and organizers have long demonstrated about carcerality transpiring well beyond the confines of prisons, jails, and po-lice, often abetted by technology.[10] As Michelle Alexander observes, "'Mass in-carceration' should be understood to encompass all versions of racial and social control wherever they can be found, including prisons, jails, schools, forced 'treatment' centers, and immigrant detention centers, as well as homes and neighborhoods converted to digital prisons."[11] Put another way, "e-carcerality" bleeds into domestic space through the digital, abetted by an arsenal of land-lord technologies and techniques.

Landlord technology is often referred to as "proptech" by the real estate industry, referencing the products, software, hardware, apps, tools, databases, assets, AI promises, and virtual platforms intended to "disrupt" the real estate industry and make it more "frictionless," "connected," and cutting-edge. Ap-plied in residential, commercial, and industrial buildings alike, proptech includes digital property management, virtual landlordism, automated payment systems, tenant screening platforms, short-term rental management, roommate-matching services, home and neighborhood surveillance tools, facial recognition and AI-enabled building access control, and more. That said, the term *proptech* itself, along with its industry-aggrandized product descriptions, is difficult to map from a tenant perspective, coded in insider argot. For instance, when interviewing the founder of the organization Women in Proptech, I was schooled that I should not be referring to those who live in homes owned by landlords as "tenants"; as in the real estate industry, the word *tenant* refers to only a renter of commercial space. I was talking about "residents," she edified, trying to be helpful. Yet by and large, those who reside in buildings owned by landlords do indeed understand themselves as tenants, as does the broader housing movement.

This discursive disjuncture inspired a research project that I have been leading in collaboration with members of the Anti-Eviction Mapping Project (of which I am a part), the OceanHill-Brownsville Alliance, (people.power. media), the AI Now Institute, and the University of Washington, where the Anti-Eviction Lab is housed. Together, we have been mapping out residential proptech systems according to tenant harms. After numerous workshops with those who understand themselves to be tenants and who had never heard of the term *proptech*, we came up with *landlord tech* to reframe the industry and signal its arbiters and benefactors. As so much landlord tech is intentionally hidden in corporate black boxes while nevertheless preying upon tenant data and lives, and as there are no industry maps of corporate deployment or associated harms, we sought to flip the gaze back on landlord tech companies themselves in order to create housing-justice-oriented knowledge. Our research now lives on our public Landlord Tech Watch web page, complete with a guide for organizing against landlord tech surveillance.[12] In what follows, I build upon the politics of this collaborative work while also exploring the industry's historic reliance on racial technocapitalism to expand its property empire.

Tenant Screening Pasts

The liberally enshrined freedom to own stolen lives and land has long been a foundational technology of landlordism, as have ongoing racial technocapitalist efforts to police, patrol, and survey who and what are considered property. The contemporary landlord tech industry embraces this, innovating new systems to accumulate land and housing by capitalizing upon tenant data. The tenant screening industry is an important contemporary node here, as it co-opts and privatizes data about tenants to sell back to landlords, so that landlords can vet who may or may not be a "problematic" tenant. Today there are over two thousand tenant screening companies in the United States alone, comprising a billion-dollar industry dedicated to algorithmically compiling, selling, and merging tenant data. Most often, housing court records, eviction data, criminal records, and credit rating scores are algorithmically (and often erroneously) scrambled and then abstracted as "thumbs up" or "thumbs down" summaries sold to landlords. Nine out of ten US landlords use some form of screening today, arbitrating whether prospective tenants can access rental housing based on screening report recommendations.[13] By updating centuries of exclusionary property practices including forced removal, racial covenant laws, redlining, segregation, and urban renewal, screening maps onto a thick palimpsest of dispossessory surveillance techniques.

While tenant screening bureaus were institutionally formalized in the mid-1970s, earlier property listings across the United States recoded these exclusionary pasts. A 1946 property advertisement in Washington, DC, for instance, implied that prospective tenants of color must be employed, of "good character," and refined.[14] In 1955, landlords in Hyde Park, Chicago, launched a nine-point code to formulate residential standards, which, while explicitly embracing interraciality, also included policies to exclude "'bad actors' or persons recently guilty of offenses including moral turpitude, or persons with an unsavory reputation or with habits offensive to their neighbors."[15] The code additionally suggested that landlords would engage in investigative work to ensure tenants maintain good credit and income levels at four times the rental price. Such practices offered landlords rather than tenants the power to define particular neighborhoods (much like homeowners' associations and tenant screening companies today), while at the same time promising tenants the possibility of housing if they could successfully perform what Emma Power and Charles Gillon describe as the "good tenant."[16] As a promissory note mired in what Lauren Berlant has called a form of "cruel optimism," such mechanisms continue to lure prospective tenants into holding on to the assurance of being securely housed if they can successfully embody racial, gendered, classed, and sexualized moral constructs.[17]

Tenant screening became institutionalized with the launch of U.D. Registry in 1975 in Southern California, which collected and compiled "unlawful detainer" data. Unlawful detainers are the summons notices served to tenants by landlords mandating that tenants either go to court to contest putative lease violations or leave their homes for good. The company realized that by commodifying these data, it could sell information to landlords interested in blacklisting "bad" tenants. Prior to U.D. Registry, eviction data were primarily confined to county and housing courts. U.D. Registry brought these data into the proprietary private realm, bearing two million records of tenant dispossession by 1985.[18] During the 1980s, a host of other screening services also emerged, compiling three categories of tenant data: public record information (particularly court disputes between landlords and tenants), financial information similar to that maintained in a traditional credit report (bank accounts, bill-paying habits, occupation, assets, and income), and "lifestyle" information culled from methods ranging from investigative research to interviews with past landlords. It is notable that the tenant screening industry emerged on the heels of civil rights protections such as the 1968 Fair Housing Act, which protects tenants from overt racial discrimination in the housing sector, and the 1970 Fair Credit Reporting Act, which aims to ensure accuracy and fairness of consumer information in the files of reporting agencies. By datafying and abstracting racially coded tenant data (in which tenants are also rendered the consumers of landlords' assets), racist decision-making was able to continue covertly. Through datafication and abstraction, then, the industry has sought to bypass anti-discrimination measures and rectify the specter of a capitalist crisis imposed by antiracist housing organizing.

Early corporate tenant screening practices were also intimately tied up in Silicon Valley developments in personal computing technology and computer banks, which facilitated inexpensive modes of data collection, transmission, maintenance, storage, and access. The Apple II computer, along with the twin floppy disc IBM PC XT, for instance, supported spreadsheet applications such as VisiCalc and SuperCalc. In 1978 Robert Benson and Raymond Biering suggested, "With the advent of the computerized consumer reporting industry, it has become possible [for landlords] to purchase a great deal of tenant information that would otherwise be too expensive or impractical to obtain."[19] Significantly, this move toward digitization transpired alongside novel utilizations of platform technologies in urban planning and "renewal" processes, including geographic information system modeling, cybernetics, and urban control rooms.[20] Brian Jefferson argues that the underlying logics of this digital transformation—one which suggested that machines can and should control poor people of color in urban space—transpired during a moment in which the private sector began driving public infrastructure development, urban operations, and ownership.[21]

Early on, concerns were raised about the possibilities of automation bolstering an already biased and privatized screening industry. In 1987 Robert Stauffer warned: "Suppose that in order to find housing in our society, you had to present yourself before a large machine. This machine, in considering your application, would review your life to determine whether you were worthy of housing."[22] This machine would have access to all public records, bill payment information, and any landlord or neighbor accounts of you having previously been "too loud," a drug user, or "simply too radical." As he worried, such omnipotent technology may very well mediate who does, and who does not, gain access to housing. This could have everything to do the automation and augmentation of already existing racial capitalist biases and moral codes. While Stauffer's technodystopic fears transpired during a time in which such a vision of automation was still science fiction, years later with new technological advances, it morphed into reality.

With the internet, the web, and the personal computer developing alongside Cold War disintegration, it became clear that commercial consumer-based tech would impel the post–Cold War era into new futures.[23] Companies such as Netscape, Cisco, Hewlett-Packard, 3Com, and Intel exploded software job growth, inciting the late 1990s dot-com boom and its corollary gentrification. This era also saw courts begin to digitize their files, leading to a surge in tenant screening. By 2003 alone, 94 percent of criminal history records (71 million records) maintained by states had become digitized.[24] But also, with the post-9/11 "war on terror" exacerbating the ongoing "war on crime," there was a bolstered market for threat filtration that screening could support. One firm alone reported having performed only three thousand annual screening checks in the year before 9/11 but 25,000 in the year following.[25] The war on terror thereby became instrumentalized through racial technocapitalist disaster logics to benefit landlord technological growth—much as COVID-19 was two decades later.

While 9/11 and the dot-com boom indelibly shaped landlord technologies, the 2008 subprime crisis and the corollary tech boom 2.0 played a role in growing the industry beyond tenant screening. The proptech investment firm MetaProp (which today manages funds for real estate investors across fifteen billion square feet of real estate assets) claims to have coined the name *proptech* in the aftermath of 2008. Yet as a timeline displayed in the firm's Manhattan office suggests, the advent of proptech likely came with the invention of the tent, back in 13,000 BCE. Next, per the framed poster, there were bricks, aqueducts, the Roman Colosseum, and—soon enough—the internet, personal computers, Amazon, WeWork, Trulia, and, of course, MetaProp. I first encountered the timeline during a visit to interview the company's cofounder, who generously autographed his coauthored book *PropTech 101: Turning Chaos into Cash*.[26] Above his signature was a personalized note encouraging me to "Embrace the #chaos!" This is one of the book's core themes, illustrated by an image of a ball of wiry chaos being funneled into a sieve with a caption about harnessing chaos to move from a place of defense to offense, from one of observation to one of joint ventures, investment, and money.

Yet neither in MetaProp's book nor on its timeline did I find mention of the subprime mortgage crisis—the event that catalyzed the protraction of landlord tech beyond tenant screening. It was then that racist risk assessment algorithms racialized Black and Latinx tenants as "high risk," with "subprime" itself a racial signifier marking those slated for dispossession.[27] In subprime aftermaths, Wall Street investment firms and trusts such as Blackstone swept in to purchase foreclosed homes at auction, employing thousands of shell limited liability companies and limited partnerships as vehicles for incorporating homes of the racially dispossessed into their booming portfolios. Investors in multifamily rental units soon followed suit, ushering in the age of the corporate landlord.[28]

The corporate use of shell companies for purchasing and often evicting tenants from buildings in order to raise rents or flip units is strategic for tax purposes but also for anonymity, with tenants today often unaware of who their landlords and evictors are.[29] In the age of corporate landlordism, it has become increasingly difficult for tenants to know who their landlord actually is or that there may be hundreds of other tenants facing eviction in buildings owned by subsidiary limited liability companies and limited partnerships of the same ownership network. Such knowledge would arguably incentivize tenant power and alliances and embolden tenants in organizing against specific evictors, which of course landlords would rather avert. In this way, while landlords desire to see and know everything about tenants through tenant screening and other technologies, they go to great lengths to prevent tenants from even knowing who they really are.

With the dawn of corporate landlords owning and managing tens of thousands of properties, a new demand was created for automated property management and administration. This saw massive investment in what Desiree Fields calls the "automated landlord," a digital surrogate for property ownership and management.[30] Landlord tech companies such as Yardi quickly began formulating tools to meet this demand, overseeing rental payments, building complaints, building access, tenant screening, and more through virtual platforms. Thus, not only did it become harder for tenants to know who their landlords were (and therefore more difficult to organize against landlord abuse), but also it became next to impossible for tenants to communicate directly with their landlords.[31] Instead, tenants became forced to communicate through often inaccessible and at times faulty apps.

Meanwhile, short-term-rental app-based platform real estate companies such as Airbnb also emerged during this era, as did a proliferation of "smart home" systems. This saw the conversion of long-term housing into short-term vacation rentals for the rich, while also incentivizing landlords to incorporate smart home "amenitization" systems into homes for wealthier tenants interested in smart appliances and concierge systems.[32] On one hand, then, digital advances offered corporate landlords a means of scaling up property management and acquisition while maintaining a veil of opacity and therefore immunity to tenant organizing. On the other, innovation offered landlords new tools with which to attract higher-paying tenants. In both cases, landlord technologies helped facilitate the automation of gentrification by capitalizing upon tenant data and housing.

Landlord technologies have continued to incorporate new techniques of prey-ing upon the datafication of tenancy while continuing to shroud landlords themselves through digital buffers. New scopic systems such as "digital door-men" have been developed to amenitize and securitize high-end buildings, attracting wealthy renters unafraid of increased surveillance, who are more worried about keeping others out than about how such systems might be used against them.[33] At the same time, low-income and affordable housing complexes have become rendered laboratories of increased tracking and monitoring tools. Biometric facial recognition systems are being developed to "catch" lease viola-tors, evict "delinquent" tenants, and raise rents. Such experimentation has been sanctioned and even encouraged by local municipalities, with New York City deputy mayor for housing and economic development Vicky Been proclaim-ing: "We've been saying to the proptech industry, 'Look you've got a resource here. Let's talk about how we can make our buildings available to test out some ideas.'"[34] And indeed, low-income tenants (rather than simply the buildings) in New York City and beyond have been widely experimented on by the industry.

For instance, in 2014, a biometric facial recognition system developed by a former Israel Defense Forces general was implemented in Knickerbocker Vil-lage towers in Manhattan's Lower East Side, home to thousands of low-income and primarily Asian American residents, none of whom consented to the de-ployment and many of whom have organized against it.[35] The same system was then rolled out in low-income housing in Harlem, with similar systems in the Bronx advertising their ability to detect illegal subletting and link to law enforcement.[36] Similar carceral biometric systems have been deployed in public housing across US cities including Cincinnati, Grand Rapids, Omaha, and others.[37] While CCTV cameras have long plagued public and low-income housing by protracting carceral scope into domestic space, the deployment of AI systems offers landlords more refined tools with which to "catch" tenants for lease violations, evict or fine residents, and raise rents. Because eviction rec-ords are then imported into the screening industry's databases, new cycles of housing precarity and carcerality become cyclically automated.[38]

Such automation coincides with novel advancements in AI systems, amounting to an era that some have likened to a new gold rush or dot-com boom; others have warned that it signals a new scale of corporatized computa-tional power dressed up in the garb of magical automation.[39] In their quest to grab data (as one form of property that abets the accumulation of another), AI systems mobilize facial recognition algorithms known to reproduce white cis heteromasculine notions of normalcy, echoing prior pseudoscientific eugenic fantasies and rendering false positives and misrecognitions.[40] Given that it is largely not white cis heteromasculine people living in public and low-income housing today, this is significant. Racial biases and failures in facial recognition systems have inspired bans in select US cities, but these do not apply to private landlords, who remain immune to regulation.

In 2018, a battle against the deployment of facial recognition in low-income housing in Brownsville, Brooklyn, brought struggles against landlord tech to the fore. Robert Nelson, owner of the 718-unit rent-stabilized Atlantic Plaza Towers (APT), decided to install StoneLock's True Frictionless™ Solution—a heat-mapping facial recognition platform. The idea was that the tenants, most of whom are Black and many of whom have been living in the building for generations, should have their key fobs replaced by StoneLock's system to track their entry into their building and thereby improve tenant safety. But for the tenants, safety was not a primary concern and rather a code word for gentrification. As Fabian Rogers, an APT tenant organizer who went on to cofound the OceanHill-Brownsville Alliance, described during a 2019 city council hearing: "I had many concerns as a tenant, and security was not one of them. It was the landlord's concern, and it was imposed on me. I already feel well enough surveilled with all the cameras and key fobs that exist. I kind of feel like a criminal even though I pay my rent on time." Or according to Tranae' Moran, another APT tenant who worked alongside Fabian in cofounding the OceanHill-Brownsville Alliance, StoneLock was intended to produce a carceral environment in the building, one that would feel like juvenile detention for her young son. Anita, another APT renter, observed, "Tenants have so many issues that need to be addressed, but now we're dealing with this. . . . So poor people like me can't live here anymore. I'm pissed at what's going on. So many people in the neighborhood are being pushed out. . . . Please consider this a tragedy waiting to happen."

Anita testified only months before the COVID-19 pandemic blanketed the globe, concentrating in the heart of her Brooklyn neighborhood. Not only did COVID-19 instigate the disproportionate death and sickness of those already bearing the brunt of systemic racism across uneven scales, but it also led to hundreds of thousands of tenants unable to pay monthly rents, disproportionately Black women.[41] The pandemic, along with the renewed uprising for Black lives, was then rendered a green light for increased technological solutionism under the auspices of virus/human tracking. In this way, Anita's "tragedy waiting to happen" marks the anticipatory temporality of racial technocapitalism and the conjoined crisis and property logics it recodes.

Looking to capitalize upon the crisis, StoneLock's CEO, for instance, sent out an email to potential customers portraying COVID-19 as "the beginning of a new way of interacting professionally. . . . Over the next few weeks we'll be reaching out to discuss the benefits of frictionless access control."[42] Meanwhile, the company BioConnect began marketing flawed surveillance systems that allegedly detect viruses and allow landlords to keep infected tenants out of buildings, pathologizing those infected. Such techno-imaginaries portray a world of interconnected surveillance, in which a person's bodily biometric identifiers and even contagions are analyzed by landlord tech surveillance companies and the property owners they serve. In the realm of tenant screening, companies such as Naborly emerged to promote tracking tenants unable to pay monthly rents. In an email to landlord subscribers, they praised their system's ability to "confidentially" snitch upon tenants and thereby assist future landlords in knowing "if a tenant has been delinquent in the past, while also helping Naborly continue to deliver the most accurate and up-to-date tenant screening services in the market."[43] For those evicted due to rental nonpayment, companies such as Civvl popped up essentially to "gigify" the eviction processes. As Civvl advertised on its website with a photo of Black men packing up boxes, "Since COVID-19 Many Americans Fell Behind in All Aspects: Be Hired as Eviction Crew." While it may not be landlord tech companies filing eviction notices or denying entry to tenants, they, like law enforcement, are embedded in a carceral network dedicated to preserving what Cheryl Harris has called the "whiteness of property"—a construct reliant upon the continuity of settler colonial and chattel slavery understandings of ownership (not to mention labor).[44] Throughout the historic trajectory of property privatization, this construct has repeatedly metastasized in times of crisis, always to reify property's whiteness.

Beyond Property

Amid the ongoing calls to abolish police, scholars such as Rinaldo Walcott have importantly clarified that such a future is not feasible without also abolishing property.[45] This intervention builds upon Karl Marx and Friedrich Engels's maxim that decommodification requires the abolition of private property but takes it further to account for the racialized and colonized lives, worlds, and data that have long been, and continue to be, fodder for property accumulation.[46] This intervention is instructive in conceptualizing an abolitionist approach to landlord technology, which encompasses calls to cancel rent, rematriate stolen land, and abolish proprietary tenant data and landlord scopic techniques. Yet such an abolition requires undoing the very structure of landlordism itself, for the conditions of a world in which property owners can possess tenants and tenant data cannot but help reproduce technologies of what

Ananya Roy describes as racial banishment, or "a form of dispossession that ensures expulsion and enacts civil and social death, and often actual death."[47]

And yet, as Roy also notes, there are other models of property beyond those of dispossession. Part of her politic of undoing property builds upon organizing practices of "dis/possessive collectivism," or a communal propertied potentiality sutured together despite contexts of racial banishment.[48] Such practices destabilize what or who, in Nicholas Blomley's words, "counts as property."[49] There are, after all, feminist reclamation projects today, such as the Moms 4 Housing struggle in Oakland, a movement led by Black unhoused mothers who reclaimed vacant housing owned by the corporate landlord Wedgewood and who then put it into the hands of a broader collective based on feminist practices of care and justice.[50] Across the Bay Area and beyond, tenant associations and unions have been growing to hold other corporate landlords such as Veritas, Lantern, Mosser, JDW Enterprises, and more accountable. There is also the Ohlone Indigenous land rematriation work of the Sogorea Te' Land Trust, which seeks to reclaim and preserve stolen land, burial sites, and culture in what is now the East Bay.[51]

Following these practices of dis/possessive collectivism, perhaps an abolitionist approach to landlord tech would not necessitate the dismantling of *all* technology involving landlords. Housing organizers and antieviction collectives have, especially during the pandemic, been designing and developing tools to shine light upon landlord ownership and eviction portfolios, revealing corporate property networks and shell companies. This includes the Anti-Eviction Mapping Project's Evictorbook, which uses a spiderweb-like graph database to bring together eviction data, corporate ownership data, and parcel ownership data in San Francisco and Oakland so that users can look up addresses or landlord shell companies and place them within a larger network of ownership and evictions.[52] As a collaboratively produced geospatial software, Evictorbook works against the racial technocapitalist logics of tenant screening, facial recognition entry systems, and eviction automation software systems by turning the gaze back upon corporate landlord structures themselves. This is especially useful in contexts of multibuilding rent strikes, tenant unions, and collective organizing. Evictorbook is one of many projects that aim to do this, and sits within a cohort of similar projects such as JustFix.nyc's Who Owns What in New York City, Strategic Actions for a Just Economy's Own It tool in Los Angeles, the Landlord Watchlist Project in Pennsylvania, Property Praxis in Detroit, and more.

Evictorbook and similar tools align with practices of what Steve Mann and Simone Browne each articulate as sousveillance, or an inverse surveillance practice used to track the tracker.[53] These projects additionally build upon the rich tradition of countermapping, which seeks to appropriate mapping technologies for anti-imperial, anti-capitalist, queer, feminist, and racial justice futures and at times undoing settler spatial imaginaries and digitalities.[54] They also converse with what Ruha Benjamin articulates as techniques of "retooling solidarity" and "reimagining technology," as well as with what Sarah Elwood outlines as practices of digital thriving that "refuse/elude hegemonic digital-social-spatial orders."[55] Landlord Tech Watch, created with the Anti-Eviction Mapping Project, the Anti-Eviction Lab, and partners has also been crafted in this spirit, as it mobilizes maps, data, and the digital to provide tenants with pertinent information for organizing against landlord tech creep, particularly in the pandemic present and amid the racist backlash of antiracist organizing via installations of Google Nest cameras and the like. Perhaps its most useful component is a contribution from the OceanHill-Brownsville Alliance (built on the APT organizing efforts against StoneLock), which focuses on how to effectively and even joyfully organize with other tenants against biometric and facial recognition systems being implemented in one's home.[56]

While engaging the digital, the Landlord Tech Watch toolkit also emphasizes the need for old-school organizing techniques such as flyering, door knocking, direct action organizing, maintaining a media presence, and strategic alliance building. In other words, while there are emancipatory possibilities for using the digital to flip the gaze back on landlords and the landlord tech industry, these retooled solidarities are only as powerful as the on-the-ground efforts of tenant organizers committed to seeing their neighbors and community members beyond datafied fragments. At the end of the day, it is grounded caregiving work, or practices in which neighbors see and work alongside other neighbors in real time, that builds momentum for the abolition of landlord technology and the racial technocapitalist logics it encodes.

NOTES

1. Kim Lyons, "Amazon's Ring Now Reportedly Partners with More than 2,000 US Police and Fire Departments," *Verge*, January 31, 2021, www.theverge.com/2021/1/31/22258856/amazon-ring-partners-police-fire-security-privacy-cameras.

2. Herring, "Complaint-Oriented Policing"; Kurwa, "Building the Digitally Gated Community."

3. Gilmore, *Golden Gulag*, 14.

4. Chikako Takeshita, "On Becoming Virushuman: Contemplating Trans-specism and Artificial Humanity during the Coronavirus Pandemic," *Foundry*, October 2020,

https://uchri.org/foundry/on-becoming-virushuman-contemplating-trans-specism-and
-artificial-humanity-during-the-coronavirus-pandemic.

5. Hall, *Policing the Crisis*.

6. Bhandar, *Colonial Lives of Property*, 6.

7. Robinson, *Black Marxism*, 66.

8. Atanasoski and Vora, *Surrogate Humanity*, 15.

9. Ananya Roy, "Undoing Property: Feminist Struggle in the Time of Abolition," *Society and Space*, May 3, 2021, www.societyandspace.org/articles/undoing-property-feminist
-struggle-in-the-time-of-abolition.

10. Benjamin, *Race after Technology*; Gilmore, *Golden Gulag*; Walcott, *On Property*.

11. Alexander, foreword to *Prison by Any Other Name*, xiii.

12. Landlord Tech Watch is available at the Anti-Eviction Lab website at www
.antievictionlab.org/landlord-tech-watch.

13. Lauren Kirchner and Matthew Goldstein, "Access Denied: How Automated Background Checks Freeze out Renters," *New York Times*, May 28, 2020, www.nytimes.com
/2020/05/28/business/renters-background-checks.html.

14. "Room Rental Service," *Washington (DC) Evening Star*, October 23, 1946, B19.

15. "Drexel Owners Set Tenant Standards," *Hyde Park Herald* (Chicago), April 6, 195.

16. Power and Gillon, "Performing the 'Good Tenant,'" 459.

17. Quote in Berlant, *Cruel Optimism*.

18. Stauffer, "Tenant Blacklisting," 241.

19. Benson and Biering, "Tenant Reports as an Invasion of Privacy," 304.

20. Mattern, *City Is Not a Computer*.

21. Jefferson, *Digitize and Punish*, 166.

22. Stauffer, "Tenant Blacklisting," 239.

23. O'Mara, *Code*; R. Walker, *Pictures of a Gone City*.

24. Oyama, "Do Not (Re)Enter," 187.

25. Ann Davis, "Firms Dig Deep into Workers' Pasts amid Post-Sept. 11 Security Anxiety,"
Wall Street Journal, March 12, 2002, www.wsj.com/articles/SB1015886922323674160.

26. Block and Aarons, *PropTech 101*.

27. Chakravartty and Silva, "Accumulation, Dispossession, and Debt."

28. On multifamily unit investment, see Ferrer, *Beyond Wall Street Landlords*.

29. McElroy, "Dis/Possessory Data Politics."

30. Fields, "Automated Landlord."

31. McElroy, "The Work of Landlord Technology."

32. Sadowski, Strengers, and Kennedy, "More Work for Big Mother."

33. McElroy, "The Work of Landlord Technology."

34. Nicholas Rizzi, "NYC Seeks Partner to Vet Proptech Tools for Its Real Estate,"
Commercial Observer, December 16, 2020, https://commercialobserver.com/2020/12
/nyc-proptech-nycha.

35. McElroy and Vergerio, "Automating Gentrification," 615–17.

36. McElroy and Vergerio, "Automating Gentrification," 615–16.

37. Douglas MacMillan, "Eyes on the Poor: Cameras, Facial Recognition Watch over Public Housing," *Washington Post*, May 16, 2023, www.washingtonpost.com/business /2023/05/16/surveillance-cameras-public-housing.

38. On CCTV, see Jefferson, *Digitize and Punish*.

39. Jurgita Lapienytė, "Is the AI Boom Just Another Gold Rush?," *Cybernews*, June 26, 2023, https://cybernews.com/tech/ai-boom-gold-rush; Kak and West, *AI Now 2023 Landscape*; Meredith Whittaker and Lucy Suchman, "The Myth of Artificial Intelligence," *The American Prospect*, December 8, 2021, https://prospect.org/culture/books /myth-of-artificial-intelligence-kissinger-schmidt-huttenlocher.

40. Chun, *Discriminating Data*.

41. Molly Solomon and Erin Baldassari, "Why Black Women Are More Likely to Face Eviction," KQED, February 21, 2022, www.kqed.org/news/11905386/why-black-women -are-more-likely-to-face-eviction.

42. Erin McElroy, Meredith Whittaker, and Genevieve Fried, "COVID-19 Crisis Capitalism Comes to Real Estate.'" *Boston Review*, April 30, 2020, https://bostonreview.net /class-inequality-science-nature/erin-mcelroy-meredith-whittaker-genevieve-fried-covid -19-crisis.

43. Didi Rankovic, "Naborly Asks Landlords to Report if Tenants Are 'Delinquent' on April Rent, to Build a Profile on Them," *Reclaim the Net* (blog), April 4, 2020, https:// reclaimthenet.org/naborly-rent-coronavirus-april.

44. Harris, "Whiteness as Property."

45. Walcott, *On Property*, 11.

46. Marx and Engels, *Communist Manifesto*, 484.

47. A. Roy, "Undoing Property."

48. A. Roy, "Dis/Possessive Collectivism," 2.

49. Blomley, *Unsettling the City*, 15.

50. Summers and Fields, "Speculative Urban Worldmaking."

51. Gould, "Ohlone Geographies."

52. McElroy, "Dis/Possessory Data Politics."

53. Browne, *Dark Matters*; Steve Mann, "Sousveillance," WearCam, 2002, http:// wearcam.org/sousveillance.htm.

54. Gieseking, "Operating Anew"; Rivera, "Undoing Settler Imaginaries."

55. Benjamin, *Race after Technology*; Elwood, "Digital Geographies, Feminist Relationality, Black and Queer Code Studies," 210.

56. Tranae' Moran and Fabian Rogers, "Organizing as Joy: An Ocean-Hill Brownsville Story," interview by J. Khadijah Abdurahman, *Logic(s)*, December 25, 2021, https://logicmag .io/beacons/organizing-as-joy-an-ocean-hill-brownsville-story-with-tranae-moran-and.

Wyze Bulb Color
Dimensions: 2.5"x 2.5"x 5.0"
Weight: 0.31 lbs
Lumens: 30 - 1100
Color Temperature: 1800K - 6500K
Color Rendering Index 90+
Power: 12 W
Working Voltage: AC 120V~ 60 Hz

Hardware
Bulb Type: LED
Bulb Shape: A19
Base Type: E26
Life Expectancy: 25,000 hours
Compatibility: Android 7.0+, iOS 14.0+
Communication Mode: 802.11 b/g/n, 2.4 GHz
Integrations: Alexa, Google Assistant, IFTTT
Operating Temperature
-4°F - 104°F (-30°C - 40°C)
Recommended for indoor use
Suitable for damp locations
Humidity Range – 0-85%

Wyze Cam v3
Color: White, with black accents

Materials
Body: Polycarbonate
Lens: Glass + plastic
Camera Dimensions: 2.05 in x 2.01 in x
2.3 in (52 mm x 51 mm x 58.5 mm)

Camera Weight: 3.5 oz (98.8 g)

Phone Compatibility
Android 7.0+
iOS 14.0+

Sensor:
CMOS Starlight Sensor
Resolution: Full HD 1920 x 1080p
Field of View: 130° diagonal

Night Vision:
IR Lights: 4194onm, 4x850nm
Day/Night Vision

Audio:
Mic: ECM
Speaker: 80db, waterproof
Simultaneous 2-way Talk
Siren

Powering methods
Power adapter: Indoor 5v/1A
Outdoor power adapter (sold separately)
Cable: 1.8 meter flat usb cable

LED Indicator
Front. Red + Blue.

Connectivity: 2.4 GHz Wi-Fi

Storage
Local Storage (microSD card required)
Supports 8 GB, 16 GB, 32 GB
microSD cards in FAT32 format
Supports up to 256 GB microSD
Cloud

Operating Temperature
-5°F - 113°F (-20°C - 45°C)

Weather Resistance
IP65 Weather resistant
Indoor/Outdoor
Certifications

FCC, IC, UC, IP65 Certified,
California Proposition 65

Languages: English
Warranty: 1-Year

Google Home Mini (2nd Gen)
Dimensions: 3.85"x 1.65"
Power cable: 1.5 m
Weight: 181 g
Colors: Charcoal

Materials
Durable fabric top made from 100%
recycled plastic bottles
External enclosure made with at
least 35% post-consumer recycled
plastic

Wireless network
802.11b/g/n/ac (2.4 GHz or 5
GHz) Wi-Fi
Bluetooth® 5.0
Chromecast built-in Power
15 W power adapter

Power adapter: 100-240 V
Speakers and Microphones
360-degree sound with 40 mm driver
3 far-field microphones
Voice Match technology

Processor
Quad-core 64-bit ARM CPU 1.4 GHz
High-performance ML engine

Sensors
Capacitive touch controls
3 far-field microphones
Ultrasound sensing
Ports: DC power jack

Google Nest Hub
Display
Smart Display: 7-inch touch screen
Resolution: 1024 x 600
Dimensions: 2.65" x 7.02"x 4.65"
Power cable: 1.5 m
Weight: 16.9 oz (480 g)
Colors: Chalk

Supported Audio Formats: HE-AAC,
LC-AAC, MP3, Vorbis, WAV (LPCM),
Opus, FLAC with support for high-
resolution streams (24-bit/96 KHz)

Wireless network
802.11b/g/n/ac (2.4 GHz or 5 GHz)
Wi-Fi for high-performance streaming
Note: WPA2-Enterprise is not
supported.

Bluetooth® 5.0
Supported Bluetooth profiles: AVRCP
controller, AVRCP target, A2DP sink,
A2DP source, GATT server, GAP

Speakers, Microphones, and Sensors
Full-range speaker for crystal clear
sound, 2 Far-field microphones
Far-field voice recognition supports
hands-free use
Voice match technology
Ultrasound sensing
Ambient EQ light sensor
Capacitive touch controls

Power: 15 W power adapter
Power adapter: 100-240 V, 50/60 Hz
Ports and Connectors: DC power
jack High-performance ML hardware
engine
Sensors: Capacitive touch controls
3 far-field microphones
Ultrasound sensing
Ports: DC power jack

Philips Hue A19 Bulb
Light
Brightness 800 Lumens
Color Temperature 2700
Beam Angle: 180°
Color Rendering Index> 80

Socket Type: E26
Dimmable: Yes
Startup Time: Instant

Power Consumption 9.5 W

Operating Temperature 14 to 113°F /
-10 to 45°C
Operating Humidity 5 to 95%

Form Factor: A19
Lifespan 25 Hours
Dimensions: 4.2 x 2.4" / 107 x 61 mm

2022-12-16 10:0[...]

FIGURE INTER2.1. *Thousand Dreams of Yamur* by Hayri Dortdivanlioglu

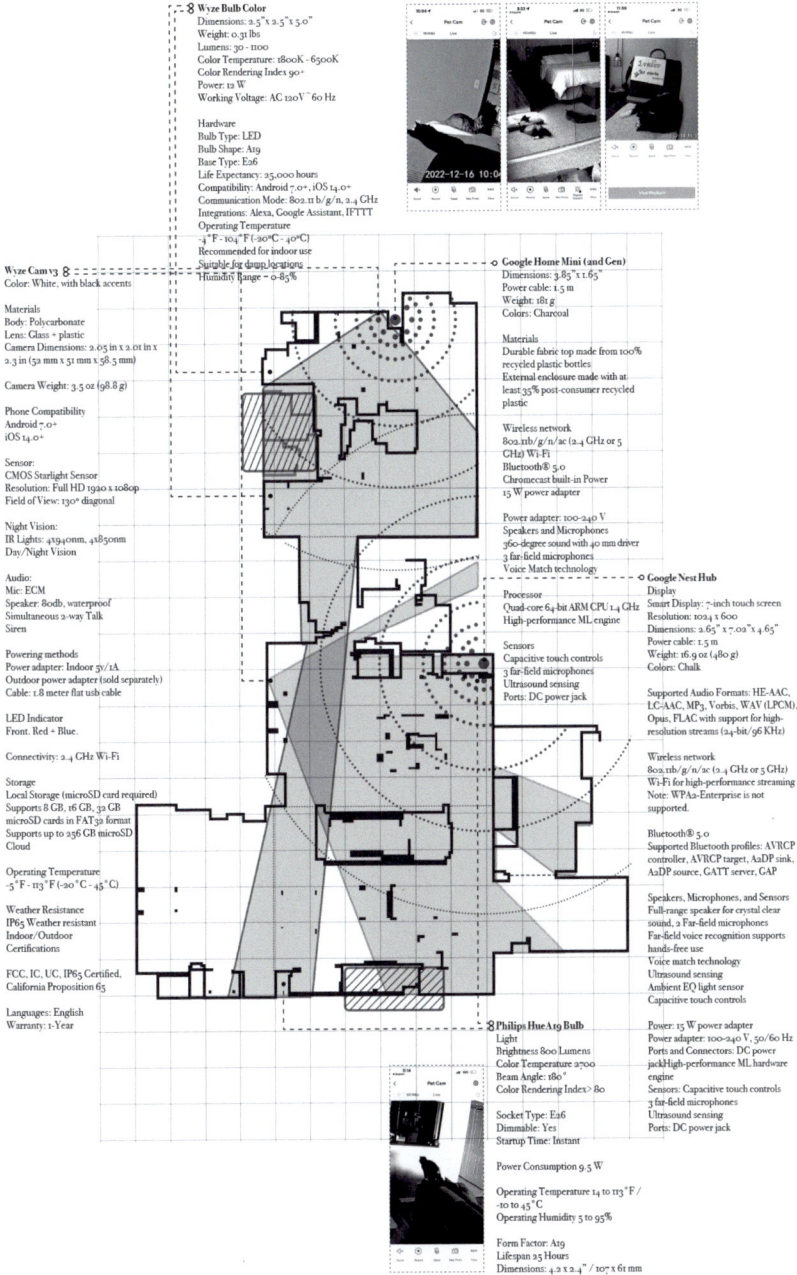

This project draws inspiration from J. G. Ballard's science fiction story "Thousand Dreams of Stellavista," where psychotropic houses, the epitome of smart homes, morph in response to the emotional and psychological states of their inhabitants. These houses, made from a bio-plastic medium, retain the psychic imprints of past occupants, leading to unforeseen and sometimes disconcerting experiences for their current dwellers. In Ballard's narrative, the protagonist, Howard Talbot, experiences haunting echoes of the house's previous tenant, Gloria Tremayne—a renowned film star infamous for her act of murdering her husband out of jealousy. A peculiar relationship forms between Howard and the memory of Gloria embodied by the psychotropic house, which eventually tries to eliminate Howard's wife, driven by jealousy. Ultimately, after being abandoned by his wife, Howard deactivates the psychotropic aspects of the house, choosing to live alone amidst the frozen memories of Gloria. While Ballard's psychotropic house surpasses our current concept of smart homes, it is undeniable that there exists a reflexive emotional bond between us and our living spaces. This emotional connection was recently exemplified when I was haunted by a recording of my recently deceased cat, Yamur. A glitch in the recording history of the pet cams I had strategically placed in my home captured Yamur in his favorite napping spot just three days after his passing. The unexpected notification, although caused by a mere glitch, stirred an intense mix of excitement, confusion, and sadness. As Benjamin Schultz-Figueroa argues, a glitch often "reveals a disconnection in humans' relationship with machines." Yet, this particular glitch in my pet cam resulted in an intense emotional bond with the technology in my hand. My home, equipped with smart technologies, and inhabited by me and my late pet, can be seen as a sentient space capable of echoing human emotions, thus highlighting the complex interplay between the human psyche and our constructed environments. This project, *Thousand Dreams of Yamur*, aims to explore these intricate relationships by visualizing the intertwined cohabitation of humans, pets, and smart home technologies. It invites contemplation of our relationship with technology within our homes and underscores the profound and often unexpected emotional connections that can be forged with our living spaces and the technology within them.

NASSIM PARVIN

NEDA ATANASOSKI

Interlude

Smart Homes

It's important we respect the privacy of your residents. Each sensor has the ability to be put into what we call "privacy mode." This means instead of the sensor showing a video feed, any people in the video are shown instead as "stick figures." —Cherry Home promotional materials

Of course, our house was automated—as all Surplus houses were required to be, by law—and the [stuffed animals cluttering the floor] could easily have been clear-floated. All I had to do was say the word and the House-Bots would emerge from their closets, their green appendages poised to help. *Clear-float now? Aren't those animals in your way?* And, *We can roll'n'clear if you'd prefer. You have a choice. You always have a choice—* the choice business being a new feature of the program. A bit of cyber-ingratiation, you might say, to balance its more habitual cyber-intimidation. *If you trip, it will be your own fault,* for example. And, *Do note that your choice is on the record. Nothing is being hidden from you. Your choice is on the record.* —Gish Jen, *The Resisters*

In March 2022, *Forbes* magazine declared the United States to be fully ensconced in an era of "intelligent, responsive homes" thanks to the ubiquity and affordability of smart home devices like speakers, thermostats, and security systems.[1] At the same time, the publication suggested that the prevalence of smart home technologies has led to consumers' increased concerns around privacy: "Having Amazon Alexa or Google Assistant may be convenient when turning off the lights or ordering in, but it also offers an open window into your home with the possibility of thousands looking in, listening from anywhere in the world. Cloud connectivity and the 'always-on' power model raise significant ethical concerns, with consumers understandably wary of their every word being recorded and transferred online. . . . Residents must also feel secure in their own homes: *recognized, not watched*. Striking this balance will be crucial for device engineers going forward."[2]

Through a series of animated art pieces that can be characterized as humorous yet creepy, Cristiana Couceiro (figs. Inter2.2, Inter2.3, and Inter2.4) visualizes this feeling of being watched by one's smart home. Colossal eyes and eyeballs invade every nook and cranny of the frame: an eyeball rolls alongside heads of broccoli, cauliflower, and bell peppers inside a refrigerator (fig. Inter2.3); a wide-open eye menacingly glares down upon a bed (fig. Inter2.4); and, most striking, Couceiro depicts proliferating eyes in the house that leave no room for human inhabitants, harkening to Alice in Wonderland, who upon consuming the wrong potion grows so large that her arms and legs seep out of a cottage doorway, chimney, and windows (fig. Inter2.2).

Couceiro's animation, in which giant eyes stand in for inter-networked home devices such as thermostats, doorbells, and speakers, were published in a 2023 *New York Times* article about people whose technologically augmented homes have seized control and taken command. The piece opens with a story about a man who was locked in his new house by his smart security system. He describes feeling like a prisoner, unable to leave until the morning.[3] In this sensationalizing anecdote, the technologies of the smart home are ultimately revealed to be creepy—threatening and repugnant.

Consumers' anxieties around loss of privacy in the digital age hinge on concerns around whether the user is in control over the smart device or vice versa. The smart home, when it robs its users and inhabitants of agency, becomes antithetical to the home's commonplace idealization as a space of comfort, safety, or innocence. Rather, it becomes strange and unfamiliar—unhomelike. Sigmund Freud's theorization of *unheimlich*, generally translated into English as "the uncanny," means unhomelike in its literal translation. In the "smart" *unhomelike* house,

FIGURE INTER2.2. Cristiana Couceiro's rendition of the struggle for control in the smart home. Animation for Anna Kodé, "Unwanted Connection: Who Has Control of Your Smart Home?," *New York Times*, February 17, 2023.

the inhabitants become commodities in the home rather than agents within it. In the expanding market for smart homes technologies become recognizable as creepy or having crept too far into our lives when they reduce users to data points to be sold—when they cease to serve the user and, rather, make the user serve the market.

On one hand, the creep of home technologies from their early inception in the 1970s to their ubiquity starting in the 2010s is a story about the ways that they are extractive and paternalistic, turning data produced within the smart home into a commodity for corporate gain. On the other hand, it is a story about users' desire for smart technologies to make their homes secure, energy efficient, and maintained by unseen but always-there *servants* who can "recognize" and "respond" to needs and desires. As people's reliance on technology during the COVID-19 pandemic (for those privileged enough to remain safely ensconced in the home) revealed, the space of the home, more than any other, has led to reflections on the ways technological platforms and objects are deemed to be either creepy or caring, either extractive or risk management tools, and either spies or loyal helpmates. The smart home hails us into a set of visual and affective practices that value a much older notion of "home

FIGURE INTER2.3. Cristiana Couceiro, animation for Anna Kodé, "Unwanted Connection: Who Has Control of Your Smart Home?," *New York Times*, February 17, 2023

improvement," including through efficiency, frugality, sustainability, and even, with the help of technology, the ability to be in multiple spaces at once. As the marketing materials for the Ring smart camera and security system tell potential customers, with Ring's technological assistance, one can be "always home."

The presumed comfort and coziness of the home alluded to by names like Nest or Ring, or the servile femininity of Alexa's voice that recalls an idealized version of a housewife at the ready to serve, surface issues of privacy and control as well as longer histories of patriarchy, whiteness, and heteronormative

FIGURE INTER2.4. Cristiana Couceiro, Animation for Anna Kodé, "Unwanted Connection: Who Has Control of Your Smart Home?," *New York Times*, February 17, 2023

nuclear family structures. We introduce the term *hometech* to contextualize the technologies that constitute the smart home within a history of the home itself as a technology of racialized and gendered property relations. As we insist, accounts that posit insidious surveillance technologies as disrupting an ideal home extend the fiction of home as a space of innocence disconnected from the operations and history of racial capitalism. Hometech produces a distinction between the inside and the outside even as it automates, manages, and makes more efficient the intimate business of social reproduction—including in the spheres of child and eldercare, health care, cleaning and shopping, to name just a few examples.[4] Yet long before the advent of hometech, the home functioned as a technology of care and reproduction predominantly maintained by women, people of color, servants, and slaves performing unwaged, underpaid, and almost always unappreciated work. Relatedly, if we understand the home as a technology that produces a set of relations, then we might apprehend the smart home as a digital augmentation of its architectural precedent, the "machine for living," heralding functionalist ideals.[5]

By contrast, growing fears that smart home technologies turn homes from "machines for living" into machines for extracting life ignore aspects of this longer history. In her widely cited book *The Age of Surveillance Capitalism*, Shoshana Zuboff argues that present-day capitalism manifests as surveillance capitalism, a mode of capitalism that "unilaterally claims all human experience as free raw material for translation into behavioral data" and, subsequently, into an expanding market in algorithmic predictions and products geared at capturing users' desires, needs, and wants—behavioral futures markets. She turns to the space of the home as the foremost example through which we can understand this emergent mode of capital accumulation.[6] Foregrounding Google's Nest thermostat—which gathers data on users' location and temperature preferences to automatically adjust to the ideal temperature, including going into an energy-efficient mode when inhabitants are away—Zuboff argues that while early 2000s visions for the smart home valued users' privacy, those days are long gone. As she explains, to use Nest, users must agree to nearly one thousand contracts.[7] This signing away of privacy to a series of contracts that an ordinary individual rarely fully reads through, let alone understands, Zuboff states, is nothing less than a "requiem for the home." In Zuboff's definition, home is not just a place but rather the space where "we know and where we are known, where we love and are beloved."[8] Now home becomes dislocated with technological disruption. Yet we might ask, Does this understanding of what destroys home reduce it to a place where one has privacy, thus erasing the complex and contradictory relations that make up homes? Moreover, does the tethering of home to privacy prevent us from reckoning with the complex racial history of the conception of privacy within the US state?

In the US legal tradition, the right to privacy has, from its inception, been tethered to racialized and gendered norms separating the private and public. The then-future Supreme Court Justice Louis Brandeis and prominent jurist Samuel Warren wrote an 1890 *Harvard Law Review* article that was the first to articulate a right to privacy in the American legal tradition. They asserted that modern media could have "corrosive effects on gendered, classed, and raced social norms." Just as today, there is growing concern about smart technologies that threaten privacy, Brandeis and Warren were concerned about the advent of instant photography that enabled the spread of tabloids and yellow journalism. The new photographic and print technologies made it possible for anyone's photograph to be taken, published, and circulated. In response, they articulated that the "right to be left alone" is critical.[9] According to Eden Osucha, Warren and Brandeis "feared that the ideal of bourgeois propriety alone was inadequate to defend 'the sacred precincts of private and domestic life' from what they framed as the 'invasion' of the domestic sphere by 'instantaneous photographs and newspaper enterprise'—that is, from the double threat of new visual technologies and related media industries."[10] Writing about the unwanted circulation and reproduction of images of white women (as opposed to caricatures of Black women like Aunt Jemima, which were normalized), Osucha contends that the "spectacles of white women in peril saturate the early discourses of media privacy, indicating how articulations of the new legal concept called on an existing cultural association between white femininity and the ideal of privacy. . . . In other words, the cultural anxieties that held unwanted media publicity to be an experience of proprietary dispossession reflect the understanding that to be subject to media publicity is to be, in effect, racialized. The racialization of cultural concepts of privacy and publicity . . . shapes Warren and Brandeis's influential redefinition of privacy as a privileged form of property."[11]

As a privileged form of property tied to gendered ideals of whiteness as innocence, the privacy ideal continues to be woven through present-day discourses on media infrastructures, including those of the smart home. Let us enter a fantasy space of a child's bedroom to observe how smart home objects reassert, rather than revolutionize, the sedimented connections between whiteness, property, privilege, and innocence. The now defunct company Quirky's version of a "smart" piggy bank was introduced in 2014. Porkfolio let children digitally see their savings, set goals, and work toward future purchases (figs. Inter2.5 and Inter 2.6). The official product description touted Porkfolio as the world's smartest piggy bank. Porkfolio wirelessly connected to an app for mobile devices that allowed users to track their balance and set financial goals from their phones. The piggy's nose lit up in celebration every time a US coin was inserted, and it could hold up to $100 in quarters.

Piggy banks are associated with the Middle English term *pygg*, used for ancient clay pots that were vessels for saving coins. As tools to teach financial literacy to children, piggy banks exploit the emotional attachment to a "cute" object as a mechanism to guard against the impulse of breaking the object that holds money. The digital augmentation of piggy banks maintains the form but is no longer dependent on the fragility of clay for its ability to safeguard the treasure inside.

FIGURE INTER2.5. Porkfolio tech specs, including piggy bank robber protection

At first glance, there is nothing that would put Porkfolio on a traditional list of creepy technologies. As a CNET review of the product stated, "If the Porkfolio's cute design doesn't immediately win you over, give it some time. The blunt-nosed guy has been sitting on my desk at work for about a week now, staring at me with dimpled eyes even as I type this sentence, and I have to admit: I've grown a little fond of the thing."[12] With its heart-tugging design, not only does Porkfolio not conjure creep in the immediate sense, but it even appears to be a harmless and frivolous object. The same review states, "The $50 Porkfolio certainly seems less like a smart home necessity than a solution in search of a problem. And yet . . . the Porkfolio takes its frivolity in stride. There's just enough charm and basic usefulness packed into this pig to keep you more or less happy with it—and it carries some unique, kid-friendly appeal, too."[13]

Porkfolio's cuteness, "charm," and frivolity all conjure a kind of technological innocence, as opposed to technological creepiness or repugnance. Thus, it is viewed as appropriate for a child's bedroom. Yet as an object that refashions the old-school piggy bank for integration in the smart home infrastructure, Porkfolio is a digital extension of property and wealth relations under capitalism, reproduced within the nuclear family. Technological innocence in the design is precisely the mechanism that obscures the cumulative creep of racial colonial ideologies in the intimate space of the home, including through the reproduction of "financial literacy" and habits of accumulation. In this sense, the in-

FIGURE INTER2.6. Porkfolio can be paired with an app.

nocence of smart piggy banks builds on a long history of what Eve Tuck and Wayne Yang have termed "settler moves to innocence"—"diversions, distractions, which relieve the settler of feelings of guilt or responsibility, and conceal the need to give up land or power or privilege."[14]

As an "innocent" technology, Porkfolio seeks to produce a virtuous and responsible individual whose goal, as a proper capitalist citizen, is to both accrue and invest their riches. Porkfolio's box reads, albeit tongue in cheek, "*I'm filthy rich.*" As C. B. MacPherson argues, "The original seventeenth-century individualism . . . [conceptualizes] the individual as essentially the proprietor of his own person or capacities, owing nothing to society for them. . . . The human essence is freedom from dependence on the wills of others, and freedom is a function of possession."[15] In the case of Porkfolio, self-possession manifests through a child's restraint, which allows them to save money. The traditional breakable materials of clay piggy banks have, after all, roots in the installation of a Protestant ethic in children based strongly on the ideals of developing self-discipline. Historically, piggy banks and money boxes, their adult counterpart, were designed to teach and reinforce such impulse control, as suggested by Jaco Zuijderduijn and Roos van Oosten based on their study of medieval financial practices in Holland. Porkfolio brings to this history a contemporary capitalist ethos, as evoked by its name's riff on the term *portfolio*.[16] It is not that capitalist values have crept into the piggy bank, which was always about teaching children the value of accumulation. Rather, with digitalization, the piggy bank is made transparent to both the child and parent so that, through the app, the process of accumulation can be tightly policed, by self and others, in unprecedented ways.

The innocence in design conceals the normalization of such relations in the digital sphere. Indeed, one of the most publicized and reviewed features of Porkfolio is its "internal accelerometer"—what Quirky dubbed "robber protection"—that sends the child a message if Porkfolio is moved or turned over. With this capacity, Porkfolio materializes a peculiar mode of seeing that may be characterized as "guarding" or "watching," often associated with surveillance practices. We could say that this charming bank is weaponized, as it comes to double as an anti-theft device. In this digital upgrade, we see a subtle but significant shift in who is being protected. If the old-school piggy bank was made to guard against a child's temptation, the new incarnation is designed to protect the wealth from others. What Porkfolio gestures toward, but doesn't make transparent, is the perpetual suspicion of service workers who may be cleaning, caretaking, or otherwise upholding (but never being part of) the white heteropatriarchal home. Porkfolio reinforces safety as the act of warding off those who may be after the child's coins. All movement is suspect. All activities need to be monitored.

Crucially, the privileges of whiteness go hand in hand with the racialized mechanisms of surveillance within heteropatriarchal family formations, of which Porkfolio is an example on the smallest of scales. Jessica Vasquez-Tokos and Priscilla Yamin have argued that "state policies have a history of denying the right to family privacy (the ability to control personal family relations and the right to be left alone from government interference) to people of color.... [The] 'racialization of privacy' ... explains that in practice the right to privacy and family recognition is a privilege of Whiteness."[17] Referencing the long history of the removal of Native American children to boarding schools as well as racial and colonial eugenic practices including forced sterilization, Vasquez-Tokos and Yamin demonstrate that racially vulnerable groups have routinely been surveilled and policed with the excuse that they pose a threat to the nation.[18]

While it is crucial to track the juridical mechanisms through which the racialization of privacy has maintained the propertied and reproductive privileges of whiteness, we underline here that extending privacy rights to everyone should not be the ultimate goal of resolving present-day fears around technological creep. As a politics, not only does privacy cleave to gendered whiteness within colonial regimes, but the struggle for privacy above other kinds of politics upholds capitalist contract and property relations as well the ideal of the self-possessed individual. It is important to recall that, after all, the treaty, the contract, and private property are the cornerstones of settler colonial relations that order the space and time of the modern nation-state. The digital architecture of the smart home seeks to make ever more efficient the reproduction of settler heteropatriarchal relations. Meanwhile, the binary classification of technologies as either surveilling us or upholding our "privacy" and "data security" affirms the home as a foremost site for the struggle for ownership over data and information—now an extension of the ownership of land and other material resources. When reduced to the surveillance/privacy binary, the home becomes a terrain of struggle (domain to be lost) for the liberal (white settler) subject. This is historically a subject who has never had to fear the loss of their autonomy and who has always held a right to property ownership, unlike women, the enslaved, and Indigenous people. The struggle to maintain command and control over data as property is about maintaining the autonomy of this subject.

In response to these anxieties, tech companies like Apple and Cherry Home have begun to increasingly market privacy as a commodity that can be bought. We can recall the iPhone ads discussed in the introduction to this book, which sell to users the ability to choose privacy in an era where all user data can be monetized. Similarly, in this interlude's epigraph, Cherry Home's marketing materials anchor their pitch for their indoor camera systems to the company's "respect" for user privacy. This marketing recalls Gish Jen's portrayal of *choices that have no impact* in the dystopic novel *The Resisters*. The novel paints a bleak picture of a near-future AutoAmerica, in which most jobs have been automated and most people are rendered "Surplus." "You always have a choice!" The irony of this repeated mantra of the automated houses is that Surplus people have no choice but to live in them. Jen describes as "the choice business . . . a new feature of the [automated home] program. . . . Do note that your choice is on the record. Nothing is being hidden from you. Your choice is on the record."[19] Indeed, transparency and choice seem to be a premium commodity touted by companies like Apple to showcase how they give their users a "choice" and opportunity to "consent" or "decline" terms of use. Recalling that in Zuboff's argument home is a complex set of affective relations and attachments that have been disturbed by hometech's drive to quantify all behaviors, emotions, and desires, we can see that the corporate response to fears about the loss of home has been to sell back to consumers a sense of autonomy and "choice" by ostensibly agreeing to recognize user data as property. Consumers' ability to see and track their data is proposed as a solution to the creep of hometech. Put otherwise, if hometech creep (which we cannot see) makes home feel "unhomelike"—making us feel watched without knowing it—then the ability to see the data collected about us promises a kind of transparency and privacy that could restore the home, or at least one version of it.

Nonetheless, given that the politics of privacy maintain the primacy of private property, individualist autonomy, and settler efficiency as the hallmarks of home rather than, for instance, collective networks of trust and care, we might ask: What kinds of relations that make up a more expansive, contradictory, and complex notion of home remain unseen in calls for technological or algorithmic transparency? In the debates around contractual agreements, data, and information, what possibilities for collectivity, community, and activism remain unnoticed? Can hometech be reimagined in a way that disrupts its norming impulses? Can it foster different sensorial and visual practices and thus different relations in spite of its intended uses? Can glitches in already-existing home technologies be put to use to push back on dominant capitalist ways of monitoring and accruing value in both critical and generative ways?

As the novel *The Resisters* unfolds, it becomes clear that the automated house itself is a character, indeed even a family member of sorts. The novel's narrator, Grant, a former professor, lives with his wife, Eleanor, a tireless civil rights lawyer who wages legal battles against AutoAmerica's policies toward the Surplus, and his daughter Gwen, a baseball prodigy. When Gwen is selected to attend an elite Netted college due to her extraordinary talents, Grant, who has spent his life attempting to subvert the surveillance of AutoAmerica's various technologies like drones and implanted microchips, finds the house attempting to comfort him, telling him that children grow up and that is the way of things. At the end of the novel, when Eleanor is arrested and ultimately murdered by the state, the automated house proclaims: "There's no forgetting what you can't forget. . . . Rage, rage against the dying of the light."[20] There can be a comfort here, in the presence of the house, that exceeds its intent and programming as a surveillance device. Gesturing at a reimagining not just of what hometech is but of the kinds of relations it can enable with human and nonhuman worlds, the house urges a rage at the order of things.

Hayri Dordivanlioglu's artwork that accompanies this interlude, *Thousand Dreams of Yamur*, also captures the possibilities that can come from a glitch or unintended function of hometech. His work, which at first glance appears to be a map of how his home is fully surveilled by Google Nest Hub and Waze cams, draws inspiration from J. G. Ballard's story "Thousand Dreams of Stellavista," where "psychotropic houses" are responsive to the emotional states and desires of their inhabitants. They also retain the memory of all past users. Dordivanlioglu's work builds on the concept of emotional bonding with technology as well as our imprints on it to visualize the "reflexive emotional bond between us and our living spaces." He describes a glitch in Yamur, the cat's, pet cam that showed his cat napping in its favorite spot just days after Yamur, the cat's, passing. The emotions he experiences, ranging from sadness to excitement, raise the possibility that the smart home can be a "sentient space capable of echoing human emotions." As he writes, emotional connections can be forged with living spaces through technology.

Centering such unexpected emotional connections and modes of living and relating, we might conclude that what we need is a requiem for the home—the kind of home that reasserts the white, nuclear settler family and capitalist property relations. This radical disruption might gesture toward alternate visions of home that engender new modes of collectivity, mutuality, trust, and struggle.

NOTES

Epigraph 1. Cherry Home (promotional materials).

Epigraph 2. Gish Jen, *Resisters*, 5.

1. Mark Lippet, "Privacy, Intelligence, Agency: Security in the Smart Home," *Forbes*, May 5, 2022, www.forbes.com/sites/forbestechcouncil/2022/05/05/privacy-intelligence -agency-security-in-the-smart-home.

2. Lippet, "Privacy, Intelligence, Agency."

3. Anna Kodé, "Unwanted Connection: Who Has Control of Your Smart Home?," *New York Times*, February 17, 2023, www.nytimes.com/2023/02/17/realestate/smart -home-devices.html.

4. Additionally, the notion that "home" and privacy are aligned belies the long history of home as a site of feminized and racialized labor. Feminist theorists including Maria Mies and Silvia Federici have argued that the space of the home has always been one of reproductive labor, where feminized, unpaid, and often unrecognized work has ensured a productive labor force. Furthermore, as Black feminists like Angela Davis have written, domestic labor (and care and reproductive work) have always been racialized and gendered. Home for some is a space of underpaid service for others. The smart home inherits these racialized and gendered notions of invisibilized racialized and gendered labor and designs them into the architecture of the new smart home that continues to reproduce and protect

the values of heteropatriarchy and whiteness. Yet in spite of the intended use and design, relations with and through technologies are always shaping and reshaping social relations.

5. Le Corbusier, *Towards a New Architecture*.

6. Zuboff, *Age of Surveillance Capitalism*, 8.

7. Zuboff, *Age of Surveillance Capitalism*, 7.

8. Zuboff, *Age of Surveillance Capitalism*, 6.

9. Osucha, "Whiteness of Privacy."

10. Osucha, "Whiteness of Privacy," 67.

11. Osucha, "Whiteness of Privacy," 73. The racialization and gendering of the right to privacy came into view in light of the *Dobbs v. Jackson* Supreme Court decision reversing women's right to reproductive choice. As a law article addressing the decision states, "State laws criminalizing abortion raise concerns about the investigation and prosecution of women seeking reproductive healthcare and about the surveillance such investigations will entail. The criminalization of abortion is not new, and the investigation of abortion crimes has always involved the surveillance of women. However, state statutes criminalizing abortion coupled with surveillance methods and technologies that did not exist pre-Roe present new and complex challenges surrounding the protection of women's privacy and liberty interests—in addition to the interests of those who may provide or help pregnant people obtain reproductive care." See Dellinger and Pell, "Bodies of Evidence."

12. Ry Crist, "Quirky Packed Its Piggy Bank with Plenty of Charm," *CNET*, September 5, 2014, www.cnet.com/reviews/quirky-porkfolio-review.

13. Crist, "Quirky."

14. Tuck and Yang, "Decolonization Is Not a Metaphor."

15. MacPherson, *Political Theory of Possessive Individualism*, 3.

16. Zuijderduijn and van Oosten, "Breaking the Piggy Bank."

17. Vasquez-Tokos and Yamin, "Racialization of Privacy," 718.

18. It is worth adding that if privacy lies on a racial and gendered continuum, we might also surmise that it can be conferred and taken away, as the 2022 overturning of *Roe v. Wade*, the US Supreme Court case which stated that women's right to an abortion rested on the right to privacy, demonstrates.

19. Jen, *Resisters*, 5.

20. Jen, *Resisters*, 298.

5

Reading the Room

Messy Contradictions in the Datafied Home

In this chapter, I draw on the artwork *LAUREN* by Lauren Lee McCarthy and the theoretical ideas of algorithmic intimacies and datafication to reveal the messy contradictions of the datafied home. What I term as the concept of algorithmic intimacies is a way to think anew about the intimate relations between subjects, algorithms, and proximities in data.[1] As the analysis aims to surface, these messy contradictions materialize in the forms of brokenness, doubt, and intimacy. Furthermore, the analysis meaningfully engages with the methodological approach of technological creep as postulated in this anthology: something almost imperceptible, a sensation gradually approaching and positioned outside of the norm. In the performance artwork *LAUREN*, the artist mimics a smart home assistant. In becoming a replicant of an artificial intelligence (AI) system, McCarthy surveils every room in the house and tries to impact the mood and feelings of her "users." *LAUREN*'s ways of perceiving, sensing, and intuitively connecting to the people taking part in the performance exemplifies techno-creep and thus opens itself up to reassessing technologically mediated relations.

The theoretical concept of algorithmic intimacies is a heuristic to challenge technodeterministic dreams of outsourcing sociopolitical issues to algorithmic systems and processes of datafication. In basic terms, *datafication* describes the process of translating subjects, objects, and practices into data points. However, the way I refer to it here points to the desire to understand all aspects of life through the aggregation and computational analysis of big amounts of data. This logic of extracting data from unprecedented sources is closely linked to capitalist incentives.[2] Mark Andrejevic refers to today's possibilities of datafication as framelessness, that is, a characteristic to understand data collection's ability "to expand indefinitely."[3] The particularity of the smart home assistant provides a case for discussing these opaque and invisible forms of datafication.

LAUREN effectively challenges technological sensibility through the ways the work mimics digital infrastructures that have already entered habitual and daily life.[4] It is a case that surfaces what usually remains invisible in the interaction with digital technologies. The reading of the artwork emphasizes that new intimacies emerge in the collection of data through the inhabitation of automated devices, sensors, interfaces, machine vision, and natural language processing.[5] With the analysis of *LAUREN*, I show the messy contradictions between the convenience in the usage of smart home devices and a nascent creepiness in the (datafied) reading of the room.

LAUREN

LAUREN is a performance artwork by artist Lauren Lee McCarthy that started in 2017 and is ongoing.[6] As part of the performance, McCarthy enters the homes of people (many of them located in the United States) to become their human intelligence home assistant *LAUREN*, following the example of smart home devices like Amazon's Alexa. The participants in the performance applied online to take part in the work and invite McCarthy into their homes.[7] The length of each performance is three days, in which McCarthy watches remotely over the home of her "users" 24/7. During this period, the artist tries to adapt to their routines, sleeping at the same time they do and observing them at every possible minute. Before the start of the actual performance, a series of custom-networked devices is installed in the home to control certain aspects. Lauren not only watches remotely but also steers light and music; she also has an overview of every room and can intervene in all aspects of her participants' home life. Fig. 5.1 shows a view from the interface panel of *LAUREN*.

FIGURE 5.1. Interface image features smart home participant collaborator Amanda McDonald Crowley, from *LAUREN* by Lauren Lee McCarthy.

McCarthy notes in an interview about her work that the home is the final frontier in which technology can enter a new space for means of surveillance, datafication, and the intimate making of relationships.[8] The home is often understood as a place where people's first socialization takes place as well as an environment in which they can really be themselves. Notions of privacy and intimacy are both commonly connotated with the home. However, given the framing of technocreep, I briefly point here to the conceptual relationship of privacy and intimacy crucial for this analysis. From the perspective of technocreep, privacy and its counterpart, surveillance, lock an understanding of technologically mediated relations in a binary. The notion of privacy is thus limited in grasping creepiness as a critical approach to analysis. Privacy is framed as the result of democratic values to protect citizens and individuals from the invasion of their personal spaces. However, the right to privacy is often reserved for privileged groups of people within societies, whereas control and surveillance of marginalized groups are legitimized—perpetuating racialized and colonial power structures.[9] A further critique of privacy as a theoretical framework is rooted in it becoming more often the center of questions around data protection. The emergence of data-driven technologies expedited the normative value of privacy as safeguard and marketing claim for tech companies. This is further discussed in the interlude "Smart Homes" in this book.

This chapter leverages the notion of intimacy as a core concept for the analysis of technologically mediated relations. Intimacy has been productive to challenge and nuance the links between closeness and power. Feminists' work on intimacy pointed out the need to pay attention to structures of dominance unfolding in spheres of the intimate.[10] In critical scholarship, intimacy is understood as a concept that has the flexibility to reinterpret systems of relationality from nonnormative perspectives and to challenge structures of dominance.[11] In other words, it is in elevating and looking at instances of the intimate that new knowledge can be generated about the sociopolitical ramifications of algorithms and datafication. *Algorithmic intimacies* is thus a framework that takes into account how deeply entangled our lives already are with algorithms and systems of artificial intelligence. In addition, it disregards any assumptions about a divide between human and nonhuman agents, as I exemplify with the case of *LAUREN* in the next section. Finally, the concept of algorithmic intimacies acknowledges the interrelated levels of the social and mediated, the technological and economic, on which intimacy and algorithms operate together.

Human in the Loop

LAUREN replicates the functionality and ubiquity of a smart home system. This is achieved not only by the artist observing the participants 24/7 but also by LAUREN interacting with her users (through text that appears on a screen) and taking control of most aspects of their homes. LAUREN supervises, exercises, and recommends actions. She dims the lights or picks music to influence the atmosphere in the home.[12] She aims to relate to and understand the needs of the participants in a different and consequently more sensitive way than smart home assistants. "I attempt to be better than an AI, because I can understand them as a person and anticipate their needs. The relationship that emerges falls in the ambiguous space between human-machine and human-human. *LAUREN* is a meditation on the smart home, the tensions between intimacy vs privacy, convenience vs agency they present, and the role of human labor in the future of automation."[13]

How does this shift to a *human* smart home intelligence system change the perspective on *artificial* intelligence systems? The strong division between what is understood as human and as artificial severely limits the analysis of the sociopolitical impact of artificial intelligence. While functioning as a form of technology herself, LAUREN embodies the in-betweenness of a technological infrastructure and the sensibility of a human. This position helps enrich the perspectives needed to effectively evaluate the many issues with algorithmic decision-making, such as nonneutral search engine results and other forms of engineered inequity.[14]

LAUREN is not just a visualization and demystification of the obfuscated technological processes of datafication, however. The artist's aim is to foster empathy and understanding for the participants during the performance—an act that algorithmically driven devices cannot achieve. The work is a reminder that human and algorithmic decision-making emerge in complex combinatory modes of being; they are highly context dependent and always include human forms of agency.[15] LAUREN is neither a neutral technology nor inherently biased but a ground from which to contextualize when and how agency is exercised by algorithmic technologies. Furthermore, if we come to understand that McCarthy's control over the participants and their homes is a result of how the users feel, act, and be, we see a more reciprocal way in which algorithmic actions come into being. To some extent, LAUREN is comparable to a set of training algorithms in which every observation and new input of information have a consequence for future interactions. In other words, this perspective fosters an understanding of algorithmic decision-making emerging in cowritten infrastructures.[16] As LAUREN, McCarthy simulates a learning algorithm through her ongoing observations and interactions with the participants in the work. This allows us to understand algorithms not as outputs based on single actions but rather as continuations and ongoing relationships with data inputs from the present and the past. The crucial difference here is that LAUREN's interactions and recommendations do not follow a preprogrammed incentive. They are based on her way of feeling and sensing, which goes way beyond the mere data collection and analysis of a smart home assistant. The artist embodies the creepiness in her gradual approach to the users, or the ways in which she learns more and more about their behaviors and needs. However, she also abstracts from previous information gathered on the participants and acts according to her human sensibility. LAUREN does not have a particular purpose other than trying to feel what her users are feeling, putting a smile on their faces, or simply wanting to surprise them. The ways in which she anticipates the needs of the participants are not restricted to correlations or patterns in data.[17] The work thus invites us to retrace how predictions and correlations of big-data analytics can be normalized and made useful for particular purposes.[18]

McCarthy's work exemplifies a defamiliarizing of technologies, which is an artistic practice used to surface overlooked elements in contemporary technology's impact.[19] In replacing a system of algorithms with a human, LAUREN and her users visualize how intimacy manifests in interactions with the (human) smart home assistant. It marks an immanent form of knowledge, a sensibility and thus intimacy between LAUREN and her users that differs from purely data-capitalist incentives to create proximities through the analysis of users' data. This intimate state also manifests in forms of dependency on one another.

Doubt

In the remote-controlled setup of the piece, the users and McCarthy enter into a relationship of reliance. The participants learn to count on LAUREN for their daily tasks and errands, such as finding their car keys. LAUREN's observation and the users' reliance on her afford a level of intimate interaction and very close observation. But it also yields feelings of doubt and uncertainty in Lauren herself, which are worth unpacking in relation to algorithmic intimacies. "'Hey LAUREN, do you remember if I took my pill?' she asks. I quickly start scanning through the footage, jumping to different moments when she might have taken it. Relying on this mix of memory and video data feels dubious and I suddenly realize there could be consequences to getting this answer wrong. I tell her I don't think so, but I'm not 100 percent sure. I'm very aware how much more confident I'd feel if I were an algorithm."[20] In this moment, LAUREN doubts her ability to have collected all relevant information in her observation of the participants. Furthermore, she is unsure whether she can retrieve the needed data from the footage quickly enough to make an informed decision. While her human memory seems flawed, the functions of data storage and immediate access seem unlimited for a computational system. LAUREN's self-reflection is a reminder of the fact that algorithmic decision-making does not add in factors such as responsibility, ethics, or the potential to create harm. The output of an algorithm is based on calculations; however, the weight and importance of an output differ greatly for the location of a participant's car keys and their intake of a needed medication.

Algorithmic systems seemingly attain unachievable accuracy in their results. However, they cannot provide purely knowledgeable outputs by default, as they heavily rely on prepared data inputs.[21] The most common algorithms today evaluate the validity of their decisions on a statistical level, even though they are implemented in social and personal contexts in which other elements of evaluation would need to be taken into consideration.[22] McCarthy's doubt about relying on video footage and her memory about an interaction with a user brings about some considerations of the ways in which algorithmic prediction is imagined. Why would her flipping through the video footage be less accurate than an algorithm? I see here the possibility of shifting the perspective on algorithms in a "diagnosis of

FIGURE 5.2. Interface image features smart home participant collaborator Amelia Winger-Bearskin, from *LAUREN* by Lauren Lee McCarthy.

disenchantment" to formulate normative assumptions about algorithms' ability to see things in their processing of data that goes beyond human capacities.[23]

For instance, the processes that machine and deep learning systems are building on rule out less likely options for decision-making.[24] There is a statistical reductionism ingrained, an aspect that Matteo Pasquinelli and Vladan Joler highlight as an "algorithmic approximation" to demystify AI and its accuracy.[25] However, vision, or the ways in which a machine "sees," is not to be confused with seeing with eyes; rather, it's seeing through the analytics of data processes.[26] These systems are trained to detect and express what they "see" in an image. Daniela Agostinho notes the questionable framing of the enhancement of vision in technology: "Machine vision occurs through data, not optical means, and through datafication, vision becomes essentially post-optical."[27] Technology, however, is based on the deciphering of single elements of information in imagery, turning calculations into single outputs.[28] If an image recognition algorithm were given the task of flipping through LAUREN's video footage to find out if the user had taken the pill, it might accomplish this task faster than a human. But an algorithm would not be aware of the potential consequences of a wrong decision. This is exactly what the doubt signals: an opportunity to question the scope of impact decisions can have.

McCarthy's reflections surface the paradoxical relationship through which we engage with algorithms today. This is the tension of being seen and the technological background of algorithms remaining unseen. How digital technologies can see us through the analyses and correlations of our various data points stands in contrast to the sophisticated processes of machine and deep learning practices that remain obfuscated. Algorithmic calculations can raise deep ethical and sociopolitical issues, as many Black scholars of science and technology have examined in various accounts.[29] There is a technological enchantment with technologies that use sophisticated algorithms. Technological enchantment is rooted in a techno-optimistic belief that AI provides magical solutions to highly complex problems. In this trope, human intelligence is not able to imitate or fully grasp AI's processes, which leads to several problems. Kate Crawford and Alexander Campolo note this in relation to deep learning:

> The discourse of exceptional, enchanted, otherworldly and superhuman intelligence shapes our understanding and expectations of deep learning systems. . . . Most important among these is that it situates deep learning applications outside of understanding, outside of regulation, outside of responsibility, even as they sit squarely within systems of capital and profit. A massively large and expensive computing infrastructure doing statistical analysis is discursively conferred the status of an enchanted object, closing them off to other forms of critique. In these contexts, the ambivalent discourse of enchanted determinism—systems that are both mystical yet profoundly accurate predictive engines—can create a blindness to forms of risk.[30]

The ways in which LAUREN feels dubious about her decisions raise questions in relation to the implementation of an algorithmic system unaware of the context in which it serves. What if all home assistant devices surfaced their doubts and insecurities as users interacted with them? Would this change something in the way users interact with them or relate to them as smart? What I aim to do with this example is argue alongside Louise Amoore that doubt is an opening, a possibility for the negotiation of responsibility and the possibility of different outcomes: "Understood thus, to be doubtful could be to experience a fullness or multiplicity of the present moment and the many ways it might unfold, such that the cruelly optimistic promises of technoscience do not cling so tightly to ideas of the optimal or the good enough output. The optimism of the algorithm is founded on optimization, that is, on its capacity to reduce the fullness of the present moment to an output to be acted on in the future."[31] The artist's doubtful response is an ethical consideration of the possibility of uncertainty, which is an impossible mode of being for an artificial intelligence system. LAUREN's reaction thus shows two forms of doubt: that is, doubt about an answer she gave to a question and doubt as a reaction to having intimate access to a participant's home. LAUREN explores the ways in which algorithmic technologies create intimacy by specific means, based on a set of assumptions: "Like an intimate companion, she helps them find their car keys and reminds them to take their medications, but they also realize that there are zones which they prefer to keep private, and that there is a line between her being 'in support' and 'in control.'"[32] LAUREN's doubt accounts for a realization of being present in the most intimate spaces during the performance. The replication of a smart home assistant thus allows one to raise significant questions as to how far the (datafied) reading of the room could and should go.[33]

Brokenness

In the final part of this chapter, I look at the LAUREN artwork through the lens of gendered and racialized framings. For this, I identify relevant aspects of technoliberalism and techno-Orientalism and put them in conversation with McCarthy's work. I amplify the idea of brokenness to argue that the ways in which LAUREN does not function like the perfect machine inherently dismantle data-capitalist beliefs about optimization and automation.

The tech industry has shown gendered biases on many accounts, sexism with regard to the design of artificial intelligence systems being just one of them.[34] With respect to the private space of the home, the industry seems to benefit from the stereotypical underpinnings of the female "role." Smart home assistants are just one example of that. The home has traditionally been framed as the place where women are "in charge" or "of service" and therefore are responsible for a load of mostly unrecognized care work. Yolande Strengers and Jenny Kennedy describe the phenomenon of the "smart wife" to account for the many female-connotated devices that have entered the home and are here to create an intimate caregiving service: "She is docile and efficient. Compliant and in control. Seductive yet shrewd. Intimate yet distant. She is ready to be played, ready to serve, and able to optimize her domain."[35] The smart assistants of today deliver this newfound paradoxical ELIZA effect, "the merging of functions and the kinds of assistance and care that a human can receive."[36]

In *Surrogate Humanity* (2019), Neda Atanasoski and Kalindi Vora criticize the sociopolitical impacts of delegating gendered and racialized work to AI and attribute these delegations to what they term *technoliberalism*—the promise that technologies will liberate humanity from dull work or other inconveniences.[37] The implication that intimate work can be automated and outsourced in different ways has a deeply dehumanizing effect on the people who usually perform these types of labor.

Together with a gendered discourse on devalorizing and outsourcing labor to the nonhuman, we need to pay attention to the techno-Orientalism framework. Techno-Orientalism describes the imaginaries around modern Asia, tracing fetishized histories of Asia as hypertechnological while at the same time intellectually primitive. As a prolongation of these myths deriving from the West, Asians are framed as dehumanized automatons and reduced to machine-like entities without critical consciousness or ethical understanding.[38] Long Bui's book *Model Machines* (2022) traces this trope of Asian bodies as work machines, extending the framework of techno-Orientalism by focusing on the particularities of the Asian American experience: "Moving out from an older colonial tradition of white Europeans siting Asia as a baffling continent of slavish lumbering masses, we find the model machine myth fully materializing out of an Anglo-American tradition in the United States. . . . The controlling image of Asians as controllable cogs marks them as not empowered intellects but encumbered bodies, a nameless sludge that is easily imposed upon. This image factor shored up the mental image of Asians as opportunistic or calculating."[39]

As a Chinese American woman, McCarthy surfaces in her performance the intersection of these prevalent structures of marginalization and dehumanization. Asian American women are particularly subjected to myths of obedience and subservience. "The U.S. techno-Orientalist imagination is thus rooted in this view of the Asian body as a form of expendable technology—a view that emerged in the discourse of early U.S. industrialization."[40] I thus argue that LAUREN challenges the imaginary of the obedient racialized servant. LAUREN is guided by her intuitive motives in her interactions with participants, which is an active act of critique directed at the technoliberalist motifs of big tech companies. Furthermore, I want to point to the notion of brokenness, the nonfunctionality that LAUREN displays as a human smart home assistant, which helps formulate another element of critique. "'I slept when they slept. I'd take my laptop with me to the bathroom. Emotionally, it was exhausting, trying to think about who they are, what they wanted.' Sometimes she was asked for things Alexa couldn't provide, or at least not yet: dating advice or a request to come over and help with a dinner party. But her 'clients' were also 'really aware that I was human, so they were patient. I was much slower.'"[41] The meaning of brokenness for the work becomes clearer through the artist's own reflections. Here, McCarthy elaborates on the emotional energy she used in the space of the performance and the exhaustion she felt when engaging continuously in caregiving tasks. Here, LAUREN debunks the myth of outsourcing intimate care work to automation.

Sarah Sharma introduces the notion of the Broken Machine to counteract the belief that the algorithmic systems in place can be fixed or made just. In this regard, Sharma notes that neither an update nor other narratives of technological solutionism can be an answer to the marginalizing, sexist, and racist biases ingrained in contemporary artificial intelligence.[42] With the notion of the Broken Machine, Sharma accounts for a feminist intervention into the dehumanizing and devaluing forms that automation creates: "New technologies will not simply redistribute power equitably within already established hierarchies of difference. The idea that they will is the stuff of utopian naivete and technological determinism that got us here to begin with. The Broken Machine should not be understood as a new technology. It is not an upgrade to an older model or a feminist design solution. . . . Like a flickering light bulb when the power is about to go out, Broken Machines signal the end of business as usual."[43]

The brokenness that Sharma describes is a form of feminist refusal to operate in line with data-capitalist principles; rather, one steps out of this norm. I understand LAUREN as such a Broken Machine, as a technology that refuses to function according to the parameters of optimization and datafication. First, the ways in which LAUREN creates intimate relationships are not for the sake of data extraction. Second, LAUREN needs breaks, is slower than a machine, and has doubts about her answers. She does not function smoothly. Furthermore, I read the work as an attempt to allude to the invisible and human-resource efforts that go into data-driven systems. LAUREN's presence renders visible the hidden structures of labor that go into the making and maintaining of digital technologies.[44] Data capitalism is characterized by the fact that every relationship between humans and nonhumans becomes part of some form of economic transaction.[45] McCarthy's work thus negotiates new ways in which interaction with AI systems is not based on such principles driven by and for capitalist incentives. LAUREN surfaces the notion of brokenness as a counterargument to the technodeterministic ideals of accurately and efficiently functioning systems. Finally, this is also in line with a refusal of a techno-Orientalist myth of Asian bodies as calculative automatons. The technological creep interferes with and actively disrupts the commonplace. LAUREN does so on multiple levels, as I have aimed to show.

Conclusion

Digital smart home assistants collect and make sense of huge amounts of data deriving from the most intimate moments. Big tech's technoliberalist beliefs are deeply rooted in gendered and racialized biases that devalue human labor and reduce intimate relationships and connections to transactions. As I have shown in this chapter, the performance artwork *LAUREN* by Lauren Lee McCarthy effectively addresses these issues. In the analysis of *LAUREN*, I demonstrated the potential of artistic practices to challenge the perception of technologies that have already entered the habitual lives of people. But *LAUREN* is also an opportunity to question data-capitalist incentives and their relations to intimacy.

NOTES

1. Wiehn, "Algorithmic Intimacies."

2. Van Dijck, "Datafication, Dataism, and Dataveillance."

3. Andrejevic, "Automating Surveillance."

4. Chun, *Updating to Remain the Same.*

5. Andrejevic, "Automating Surveillance."

6. "LAUREN," Lauren Lee McCarthy (website), accessed January 9, 2023, https://lauren-mccarthy.com/LAUREN.

7. At get-lauren.com, anyone can apply with their name, their email address, and a short video explaining why they want to try LAUREN. For a brief period, McCarthy was also inviting participants to her home in Los Angeles to take part in the artwork. Get Lauren, accessed January 9, 2023, https://get-lauren.com.

8. Lauren McCarthy, "Want a Smart Home Assistant? Invite This Artist to Watch You for Three Days Instead," interview by Hrag Vartanian, *Hyperallergic*, December 25, 2017, https://hyperallergic.com/417839/lauren-mccarthy-smart-home-assistant-interview.

9. Browne, *Dark Matters*; Vasquez-Tokos and Yamin, "Racialization of Privacy." In Denmark, the place I write from, invasions of privacy were part of colonial and racialized practices on the Greenlandic population. In 1951, the Danish state took twenty-two Inuit children from their homes and families in Greenland and brought them to Denmark. This "experiment" forcefully assimilated Greenlandic children into Danish schools and foster families in order to make them model Danish citizens. The children were kept away from their families in Greenland for at least one and a half years, with dramatic consequences for their personal lives. Six of the children were adopted by Danish families and never returned to Greenland. Thiesen, *Greenland's Stolen Indigenous Children.*

10. Antwi et al., "Postcolonial Intimacies."

11. Stoler, "Intimidations of Empire."

12. "LAUREN."

13. Lauren McCarthy, "Feeling at Home: Between AI and Human," *Immerse*, January 8, 2018, https://immerse.news/feeling-at-home-between-human-and-ai-6047561e7f04.

14. Noble, *Algorithms of Oppression*; Benjamin, *Race after Technology.*

15. Bucher, *If . . . Then*; Roberts, "Your AI Is Human."

16. Amoore, *Cloud Ethics.*

17. Video testimonial in which participants and McCarthy elaborate on their experience is in "LAUREN."

18. Thylstrup et al., *Uncertain Archives.*

19. Stark and Crawford, "Work of Art in the Age of Artificial Intelligence."

20. McCarthy, "Feeling at Home."

21. Ribes and Jackson, "Data Bite Man."

22. Amoore, *Cloud Ethics*; Gillespie, "Relevance of Algorithms."

23. Campolo and Crawford, "Enchanted Determinism," 5.

24. Deep learning is a subcategory of artificial intelligence, using neural networks to be applied on usually very complex and large data sets. Deep learning is, e.g., used to analyze natural language in online conversations, speech recognition, and facial recognition systems. Kelleher, *Deep Learning*.

25. Matteo Pasquinelli and Vladan Joler, "The Nooscope Manifested: Artificial Intelligence as Instrument of Knowledge Extractivism," KIM HfG Karlsruhe and Share Lab, May 1, 2020, http://nooscope.ai.

26. Agostinho, "Optical Unconscious of Big Data."

27. Agostinho, "Optical Unconscious of Big Data," 2.

28. Amoore, *Cloud Ethics*.

29. Benjamin, *Race after Technology*; Buolamwini and Gebru, "Gender Shades"; Noble, *Algorithms of Oppression*.

30. Campolo and Crawford, "Enchanted Determinism," 9.

31. Amoore, *Cloud Ethics*, 146–47.

32. Fedorova, *Tactics of Interfacing*, 4.

33. Dominic Rushe, "Let Me into Your Home: Artist Lauren McCarthy on Becoming Alexa for a Day," *Guardian*, May 14, 2019, www.theguardian.com/artanddesign/2019/may/14/artist-lauren-mccarthy-becoming-alexa-for-a-day-ai-more-than-human.

34. Buolamwini and Gebru, "Gender Shades"; D'Ignazio and Klein, *Data Feminism*; Rhee, *Robotic Imaginary*.

35. Strengers and Kennedy, *Smart Wife*, 13.

36. Joseph Weizenbaum's ELIZA, developed at MIT in the 1960s–70s was an early attempt to create a computer system capable of intelligent conversations with humans. ELIZA was introduced as an AI therapist, a computer program that conducted conversations with humans through a script that was linked to a database. Even more so, the term *ELIZA effect* was coined to portray intelligence and humanlike behavior and define artificial intelligence systems. Dillon, "Eliza Effect and Its Dangers." Quote in Fedorova, *Tactics of Interfacing*, 158.

37. However, this promise masks the fact that enabling digital infrastructures is heavily dependent on the invisible labor of marginalized subjects in data capitalism. Atanasoski and Vora, *Surrogate Humanity*.

38. Roh, Huang, and Niu, *Techno-Orientalism*.

39. Bui, *Model Machines*, 5.

40. Roh, Huang, and Niu, *Techno-Orientalism*, 11.

41. Rushe, "Let Me into Your Home."

42. Sharma, "Manifesto for the Broken Machine."

43. Sharma, "Manifesto for the Broken Machine," 177.

44. Atanasoski and Vora, *Surrogate Humanity*.

45. S. West, "Data Capitalism."

6

Surveillance Vigilantes

Property, Porch Pirates, and Paranoia on Nextdoor

Since 2014, Kellie has posted about the goings-on of her middle-upper-class Houston neighborhood online. Kellie is a middle-aged white woman, and her posts range in focus from tips for preparing for upcoming winter freezes to lost cats. Her most frequently engaged posts concern potential neighborhood crime. Some of the posts read like a police file in the making: "Person 1. Hair: Tan Hat. Top: Dark shirt, tan jacket. Bottom: Dark pants. Age: 40s? Sex: Female. Race: Black. Physical Build: Thin." Continuing with a warning, the post reads, "This woman was trying to sneak into my backyard as I pulled out of my driveway. We exchanged some words. Quite scary. Does anyone recognize her?" Kellie writes about a Black woman walking in the neighborhood and hopes to warn her neighbors of the supposed outsider's presence. Kellie is one of many concerned citizens and avid users of Nextdoor who use the hyperlocalized social media network as a place to vent frustrations, breed paranoia, and assert property ownership.[1]

Nextdoor has become notorious as an outlet for racial profiling through informal networks of individuals who surveil their neighborhoods in their free time.[2] Despite Nextdoor's attempts to curb racism on the platform, problematic public posts about neighborhood goings-on litter the site and create larger-scale debates about who deserves access to public and private property. There's much to be said about the effects of digital redlining and neighborhood demarcations and the role of Nextdoor in the ways it energizes the fear of the "other" and surveils Blackness.[3]

While the platform does not enable sharing beyond users' current and adjacent neighborhoods, the site facilitates a hyperlocalized mass sharing capability that reaches far beyond the local neighborhood organization's reach. Users illustrate their everyday observations in addition to expressing their technological obsession with video recordings of mundane and extraordinary moments of city life. Nextdoor posts are windows into the neighborhood moments of a city's home life and are at the core of this chapter, particularly surveillance footage posts that people share from their home cameras.

I ethnographically analyze Nextdoor posts in Houston in combination with a historical analysis of patents around hometech, following what the editors of this book describe as smart home technologies that have become creepy by reducing users to data points to be sold. Home security cameras linked to the cloud and law enforcement services have certainly become hometech. The creep of Amazon's Ring into daily life has reduced footage of routine tasks, like package deliveries, into data points to be sold. The overlap between Amazon providing delivery services and offering monitoring solutions to prevent package theft highlights the feedback loop within home technology. Surveillance cameras become hometech when users upload photos and video footage to Nextdoor, a "free" service. Like most nonpaid social media services, the Nextdoor platform uses these posts to fuel advertisements, and the advertisements fuel users' obsession with the latest technological innovations around hometech. While the desire to make homes more secure is a real concern, as shown by my ethnographic interviews with Nextdoor users, the users are less concerned with possibilities for corporate gain and collective control through hometech. Placing Nextdoor posts and hometech patents in conversation with one another allows this chapter to ask how the past can help us think critically about the present. The home surveillance camera is at once creepy and caring, extractive and a risk management tool, and a spy and a loyal helpmate in the home.

A focus on posts in which residents discuss security measures to protect their home renders a necessary conversation around conceptions of the home, property, privacy, and race. As Cheryl Harris argues, racial identity and property are deeply interrelated concepts that share critical characteristics like the right to use, reputation and status, and the right to exclude.[4] Forming the historical basis for the merger of whiteness with property, the logics of slavery viewed human life as property through the lives of Black enslaved people. An additional history of property must be written linking the home, privacy, and racial conceptions of the human. Thus, legal and political narratives that equated common-law concepts of property with civilized life were coupled with a belief in certain people's inherent superiority. Property ownership remains contingent on "property logics" that "cast certain groups of people, ways of living, producing, and relating to land as having value worthy of legal protection and force."[5] I agree with Harris that whiteness has come to have value as a property in itself, and I show how Nextdoor users protect property as whiteness, as well as whiteness as property.[6] As "a racial regime of ownership," debates around privacy through Nextdoor posts help renew conversations around privacy and help property ownership be seen as a superior form of being and living.[7] Given Harris's definition of whiteness as property, this chapter asks, How does racial capitalism and its history in the United States produce a racialized relationship to home and property?

Following this book's introduction, this chapter shows what gets forgotten when technological creep is seen as unprecedented, new, and uniquely threatening. By focusing on the similarities and differences between the first patent for home security and more recent patents, I dwell on the slowness of home-tech creep to expose "how racialized, gendered and colonial power relations are engrained in (and have crept into, imperceptibly to most) and reproduced by present-day technological use and design."[8] As such, this chapter excavates the social formations constituted by surveillance vigilantes and the technocreep of surveillance cameras by examining the development of patents around home security devices and the use of these devices by Nextdoor users. This vigilantism is read through the ways in which the security state imbricates feminist, racialized, and gendered subjects to reorganize fundamental assumptions about property.[9] In the name of security, surveillance vigilantes attempt to protect their property and their neighborhoods.[10]

Home Surveillance Systems

Surveillance vigilantes' use of hometech is only possible through the slow technocreep of home surveillance innovations over the last sixty years. Who needs to think about securing their home, particularly for those whose home was never a secure concept? In 1966, Marie Van Brittan Brown, a Black nurse from New York, invented the home security system out of a need to secure her own community in Queens in the 1960s. As a Black woman working outside the home, Brown was concerned about the safety of her home and invented a network of connected devices that is now known as the home security system. Surveillance studies scholar Simone Browne asks what it means for a Black woman to center these concerns at a time when police response was quite lax.[11] In conversation with feminist surveillance studies, I ask whether Brown's identity as a Black woman makes her version of home security different from contemporary forms of hometech patents. A recent *Wired* article poses an important question about the link between this Black inventor and anti-Black iterations of her device.[12] If systems of surveillance are always already tied to anti-Blackness, as Browne argues, then the intentions of a Black woman inventor may not excuse the ways home surveillance has been used to demarcate otherness.[13] Though Brown invented home security systems in the 1960s, the combination of her invention with the monetizing forces of selling data points rendered from the cameras' outputs was fifty years away. The history of hometech patents shows how the many ways of thinking about home security change over time. For the original inventor, home was never a secure concept because of the fraught history of whiteness as property. During the period when later inventions of home surveillance devices like doorbell cameras (such as the Amazon Ring) emerged, home security became more of a concern because major companies like Amazon purchased and marketed hometech.

Though a report in the *New York Times* gave most of the credit for the invention to Brown's husband, Albert, the US patent office acknowledged Brown as its primary creator in 1969. The system made use of existing closed-circuit television technology developed during World War II. Her invention created a closed-circuit television system for surveillance known today as CCTV. While Brown never financially benefited from her invention, the tech world does provide it with some recognition. Doorbell detector patents from 2015 and 2017 reference dozens of other patents, but Brown's patent is not referenced, which continues to invisibilize Black intellectual labor.

Born and raised in Queens, Marie Van Brittan Brown shared an apartment with her husband, Albert, an electrician. She worked long, unpredictable hours as a nurse, so her home was often unoccupied. The home security technology she invented included three peepholes in her front door with an attached camera that could move between them (see figs. 6.1, 6.2, and 6.3).[14] Surveillance scholars like Browne aptly refer to this intricate system as "1966 CCTV," an all-encompassing system of surveillance.[15] The patent is more than just a reference for a security system. The patent shows a person resting in their bedroom,

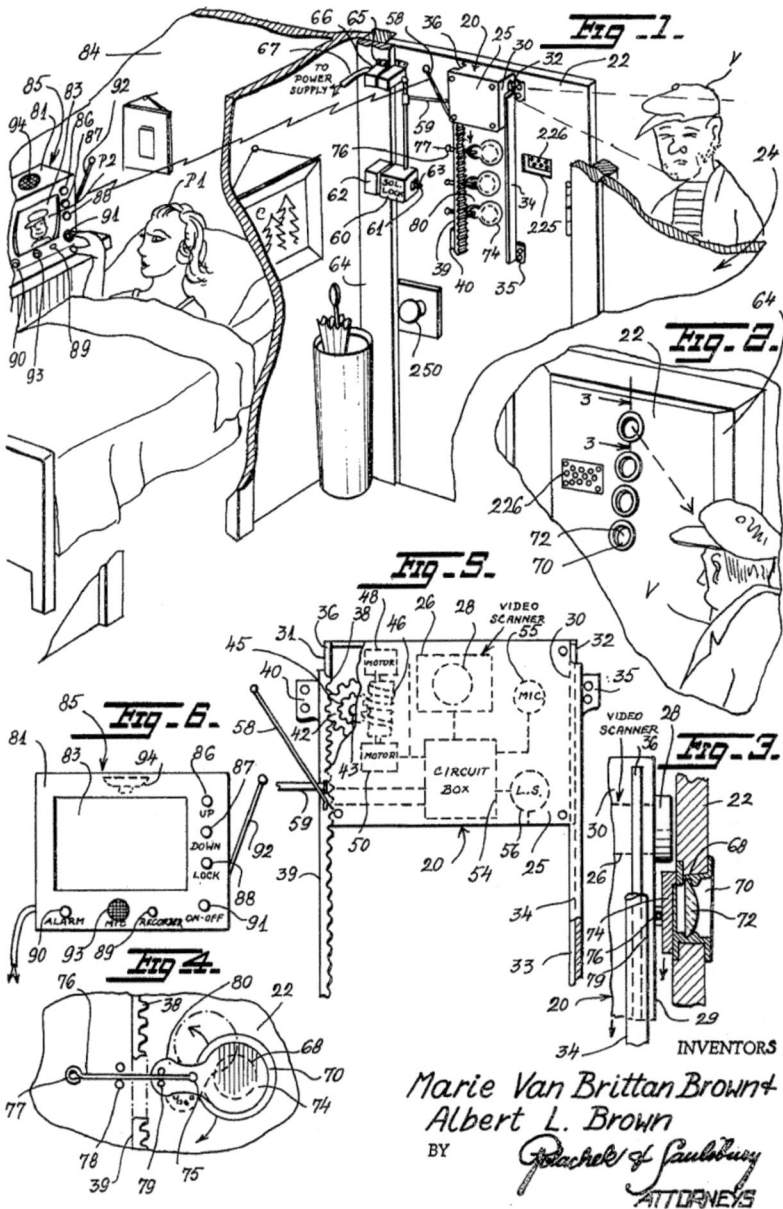

FIGURE 6.1. A woman lies in bed as a man approaches her door. Marie Van Brittan Brown and Albert L. Brown's home security patent depicts how someone could use each option of the home surveillance system while lying comfortably in bed.

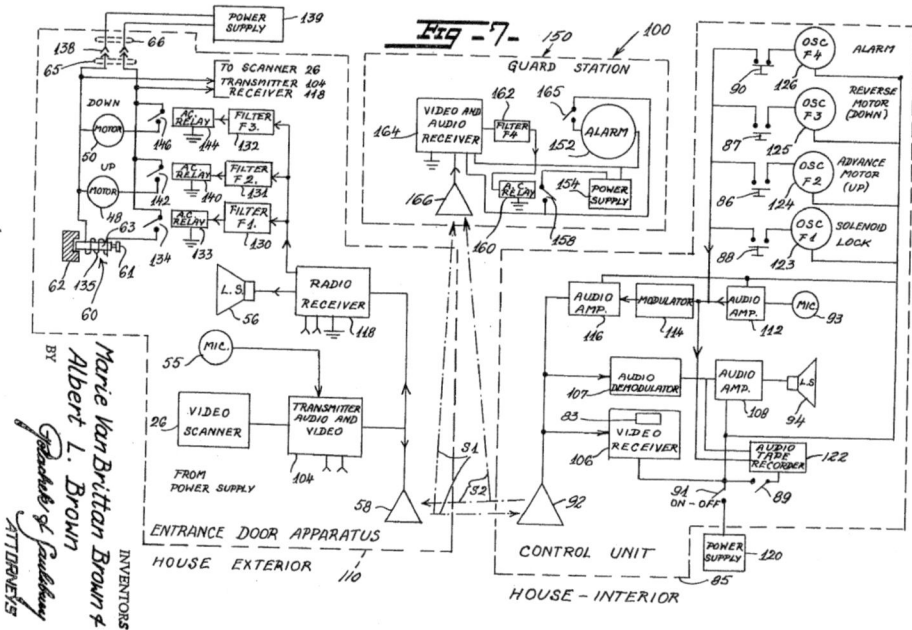

FIGURE 6.2. Diagram of the inner workings of the closed-circuit home security system from the Browns' patent

monitoring a screen projecting an image of the front-door area. In other words, the invention allowed for the projection of images across time and space to facilitate vigilance of one's property even while resting. This desire for usability is mirrored in contemporary patents for what would become the Ring home security and smart home company owned by Amazon, Bot Home Automation, Inc.[16] The ability to see outside one's property via a smartphone application is one of the main features of Ring's patent (see figs. 6.4 and 6.5). Overall, Brown's 1969 patent depicts a domestic space full of wonder and detail, with an umbrella holder and art on the walls. The mixture of domesticity and technological advancement is striking for its ability to envision safety and protection within a familiar space.[17]

The domesticity pictured in Brown's patent is one of the fictions that contemporary home security systems aim to protect. The 2015 SkyBell patents for doorbell communications systems portray a form of domesticity safe from outside forces. During the daytime (indicated by clip-art sun and clouds in the patent), a visitor receives a message from the homeowner through the doorbell: "Welcome to our humble abode" or "Welcome. We'll be right there" (figs. 6.6 and 6.7).[18] The use of *visitor* in the patent indicates a more neutral tone than the language used

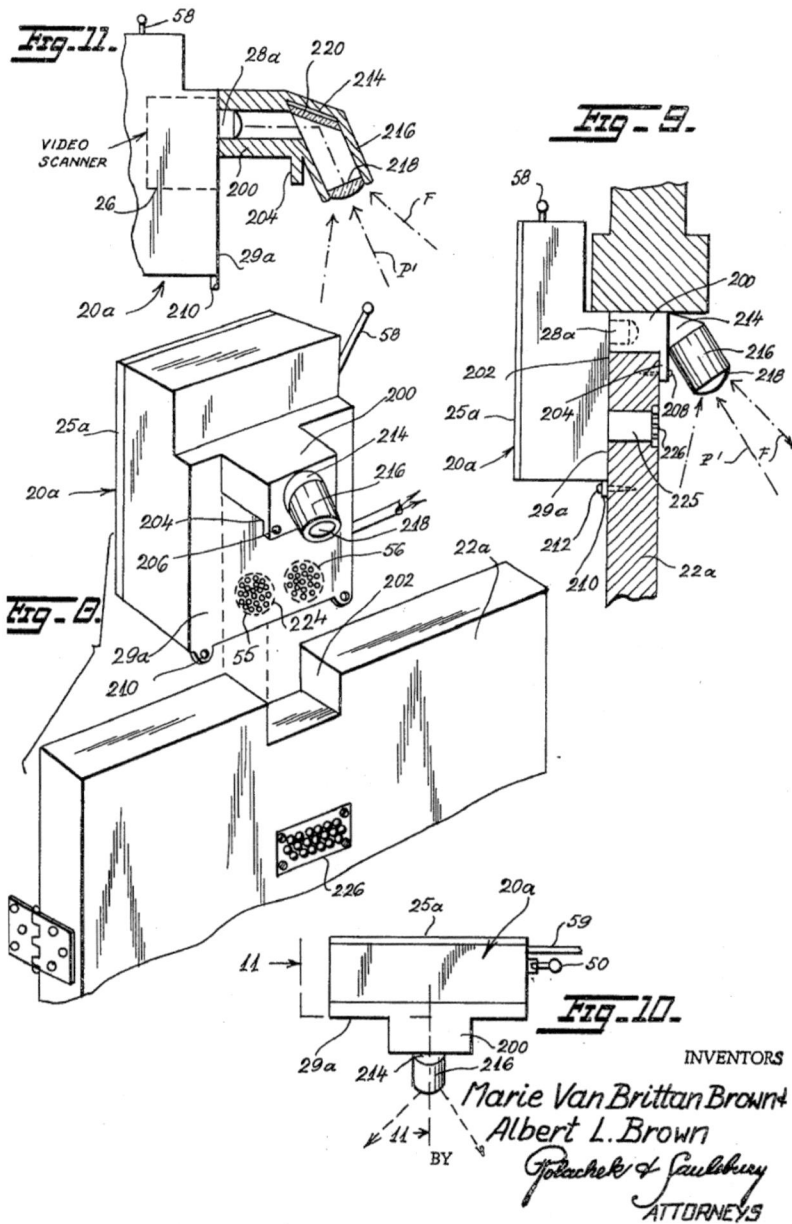

FIGURE 6.3. The Browns' home security patent depicted the possible angles for the front-door camera. Four figures show the different angles and options for the cameras.

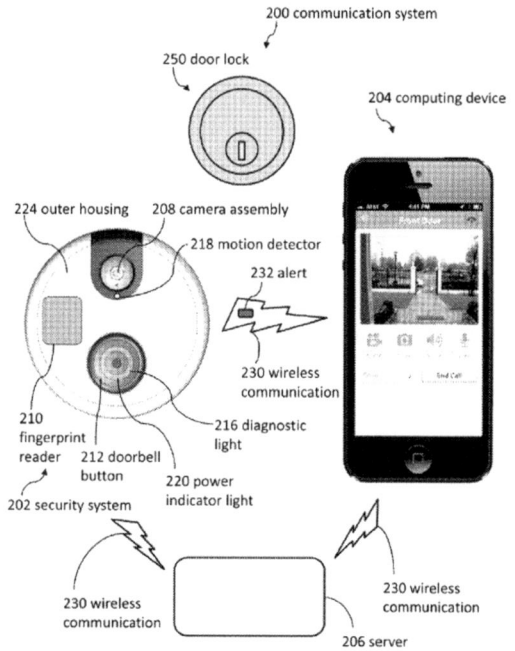

FIGURE 6.4. Doorbell surveillance patent representing how a doorbell camera and phone application work together

200 communication system
250 door lock
204 computing device
224 outer housing
208 camera assembly
218 motion detector
232 alert
230 wireless communication
216 diagnostic light
210 fingerprint reader
212 doorbell button
220 power indicator light
202 security system
230 wireless communication
206 server
230 wireless communication

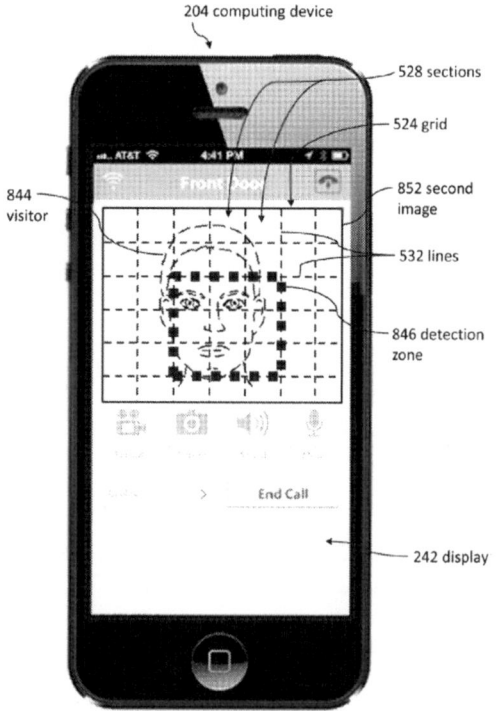

FIGURE 6.5. Doorbell surveillance patent depicting how faces are analyzed through a front-door camera

204 computing device
528 sections
524 grid
852 second image
532 lines
846 detection zone
242 display
844 visitor

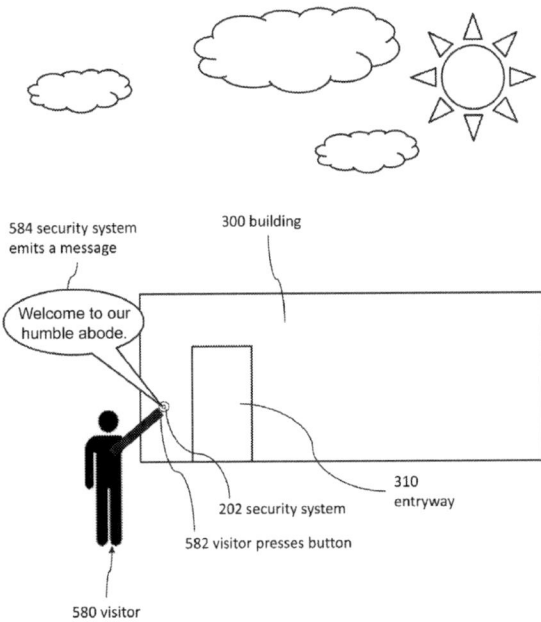

FIGURE 6.6. A doorbell surveillance patent illustrates how the person inside the home can welcome their guest through a microphone with the message "Welcome to our humble abode."

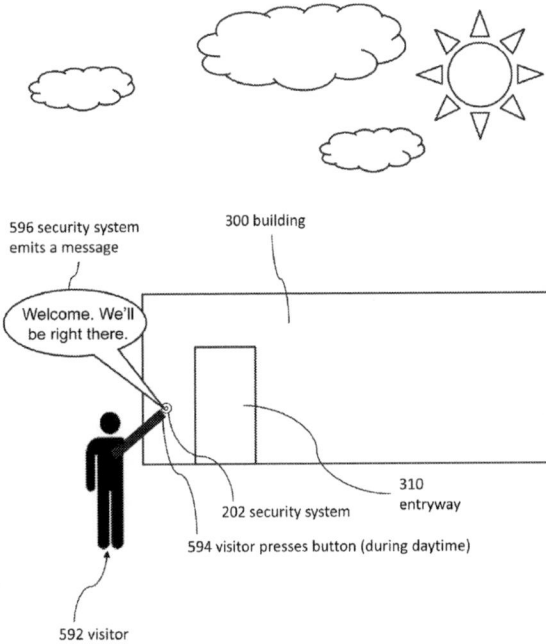

FIGURE 6.7. A doorbell surveillance patent portrays how the person inside the home can welcome their guest through the microphone by saying, "Welcome. We'll be right there."

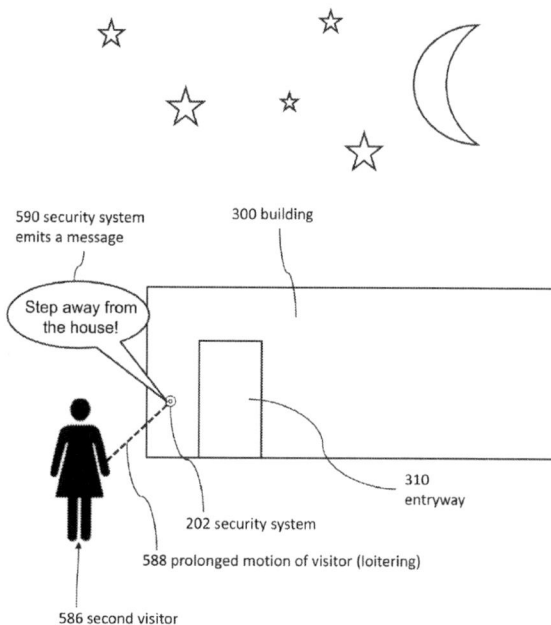

FIGURE 6.8. A doorbell surveillance patent depicts how someone inside the house can convey the message "Step away from the house!" to a visitor through a microphone.

590 security system emits a message

300 building

Step away from the house!

310 entryway

202 security system

588 prolonged motion of visitor (loitering)

586 second visitor

in the company's other home security patents. At nighttime (indicated in the patent by the clip-art moon and stars), however, a presumably unexpected visitor hears the message "Step away from the house!" through the doorbell (fig. 6.8). The change from day to night renders a different response to a person approaching the door. The change in use of the security system also changes according to the visitor's gendered appearance, as the "Step away from the house!" message is addressed to a figure in a dress in the patent. To recognize a person on private property, the doorbell "visitor detection system" combines three features: camera, motion detector, and infrared detector.[19] The invention focuses on detecting visitors outside the home at the primary entry point of the front door and relies on visual and audio indicators like knocking or talking. While the doorbell focuses on the outside of the home, the focus is still on the owners' property.

Almost fifty years earlier, Brown's patent invented the two-way radio system used in the doorbell communication system. Brown's patent depicted monitors installed throughout the premises, so she could keep an eye on whoever was outside and even speak with them via a two-way radio system. The monitors included controls to move the camera and projected images. A remote control allowed her to lock and unlock the door from a safe distance, and a panic button on the remote included a speed dial to the police. The collaboration of and reliance on police is a striking one given the history of policing and Blackness. It is also striking in context of Simone Browne's analysis of the patent through the lens of community response to the states' failure to support and protect Black communities.[20] If the original home surveillance system is "a kind of abolitionist technology," then how should present-day changes to hometech be understood?[21] The motivation behind a venture matters, and technologies can be mimicked by people who may or may not have an abolitionist commitment.[22]

With the original patent, Brown cultivated a vision of the home facilitated by a nuanced understanding of technology's ability to provide safety. The biggest difference between it and the several doorbell hometech patents of the mid-2000s is the intended use. Brown's patent and SkyBell's patent only show individual use of the technology, that is, watching one's property for the sake of individual security. In contrast, SkyBell's competitor, Bot Home owned by Amazon Ring, extends their patent and advertisements to the neighborhood's safety (figs. 6.9 and 6.10). Ring's advertisements focus on the presence of strangers and promote the utilitarian aspects of their doorbell hometech with the tagline "Always Home," extending protection to the neighborhood through their social media network application.[23] Brown's patent also focused on home security inside and outside the home, while current-day patents focus exclusively on the inside looking out. All of the patents focus on the protection of property, but two of the patents emphasize individual use of technology. The Bot Home patent, which became the basis for Amazon Ring products, instead emphasizes the community use of technology, thereby linking the product to hometech. Technocreep as theorized in the introduction to this book acknowledges this complexity of changes from individual to community use of technology. The original patent was developed before technologies such as Nextdoor, which in turn shaped that initial invention into a powerful surveillance apparatus reinforcing existing power differentials around class, race, gender, and housing status. An alternative version of home security could have built upon communal and collective ways of conceptualizing communities and collectives.

Nextdoor

In a different conception of the home, Nextdoor users cultivate a homeplace by attempting to protect property and dissuading folks of marginalized identities from entering public neighborhoods. Throughout Nextdoor posts there is an implicit "sense of loss" related to the "loss of control over one's property."[24] The platform itself demarcates who is and is not in a neighborhood. Nextdoor users can only access the forum for the neighborhood they live in and adjacent ones. Membership is confirmed by neighbors or by receiving a postcard at your address with a log-in code. In other words, membership in a community requires

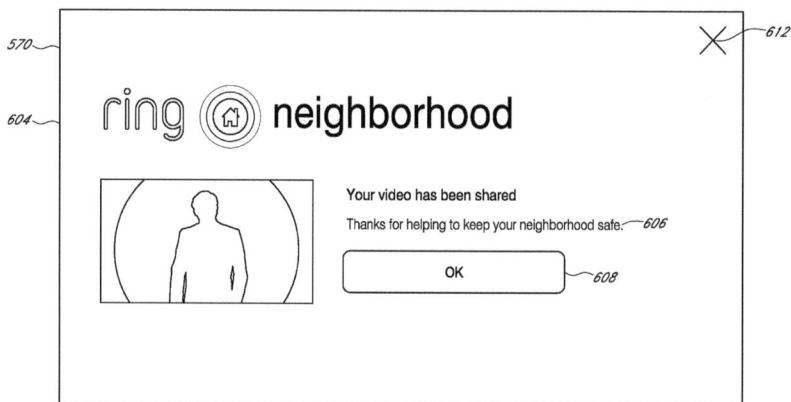

FIGURE 6.9. Bot Home Ring patent shows what kinds of notifications the phone application can show its users to upload their photos and videos to social media.

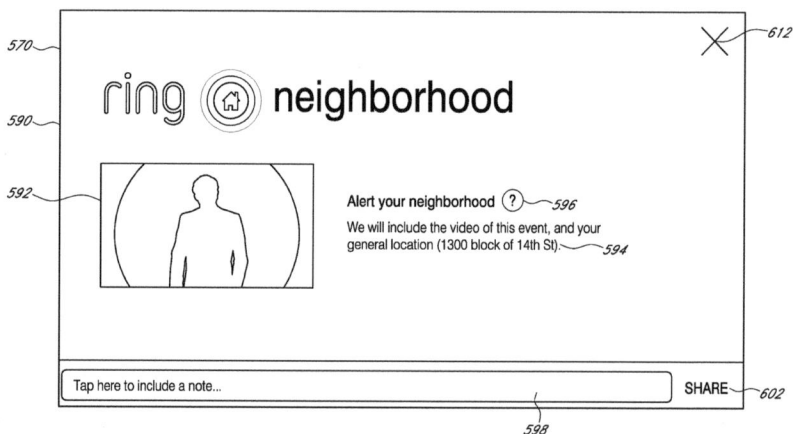

FIGURE 6.10. Bot Home Ring patent shows how a user can alert their neighborhood with security footage.

knowledge of that community either through current contacts or by living in the physical space. Embedded within Nextdoor's structure is an anti-unhoused bias. Once a new member joins a neighborhood forum, they have access to a home page that resembles Facebook in structure. People can follow certain topics of conversation ranging from foster kittens and dogs ready for adoption to concerns about power and internet outages and a buy/sell community page.

Posts are accessible only to other members of the community. Nextdoor presents itself as "a trusted, secure network of local people, rather than strangers such as those found on Craigslist."[25] The Nextdoor logo markets the platform as "the private social network for your neighborhood." The idea of a public social market that can be marketed as private promises a sense of seclusion to its users. The logo pledges privacy for your neighborhood conversations by using the familiar language of other social media networks that use algorithms that benefit posts with the greatest number of likes. In other words, users can have a sense of privacy wrapped in a familiar message of sociality. Within this familiar promise of privacy on a social media network, Nextdoor cultivates a particular sense of a neighborhood, one where a community can come together to protect property.

As geographically localized communities, neighborhoods are often social communities with considerable face-to-face interaction among members. Nextdoor aspires to the fictitious ideal of the midcentury American neighborhood as an ideal space worthy of upholding through manufactured communities, much in the way that Setha Low writes about how gated communities maintain whiteness.[26] Through a common bond based on location, residents in neighborhoods seek to realize common values, socialize youth, and maintain effective social control. One particular form of the neighborhood is the gated community, a neighborhood encircled by gates and fences, often including security restrictions for entry into the neighborhood. Gated communities can choose to implement draconian restrictions to maintain their feelings of safety and privilege.[27] Gating is only one example of this new form of social ordering that regulates and patrols the unhoused and the urban poor. While gated communities facilitate a trade of constitutional rights for insulation from socioeconomic problems and assurances of property values, Nextdoor facilitates a similar space by providing a space to protect property without the same organizing and infrastructure.

No longer are fences and gates the only things demarcating neighborhoods. Surveillance cameras and neighborhood gossip on Nextdoor also demarcate the boundaries between rich and poor, housed and unhoused, white and Black, public and private. Also relevant here is Trayvon Martin's death at the hands of a neighborhood watch captain, George Zimmerman. While Nextdoor was not a factor in Martin's death, the sense of vigilante surveillance has fueled incidents beyond Nextdoor. This sense of vigilantism extends to moments like the unnecessarily aggressive treatment of a young Black kid at a McKinney, Texas, neighborhood pool.[28] In other words, the demarcations of neighborhoods in the physical and digital space matter. More broadly, surveillance techniques that occur within Nextdoor facilitate a culture of neighbors' constant gaze, reminiscent of Jane Jacobs's "eyes on the street."[29] However, Jacobs's concept builds on an idea of community and collectivity in a way that Nextdoor does not. Nextdoor facilitates the surveillance vigilantes. Like the difference in patents discussed above, the intentions for community building change the understanding of the technology.

On Nextdoor, there are countless posts from concerned neighbors about packages, plants, or personal items being stolen from their property. The popularity of the term *porch pirate* (i.e., people who take packages from people's porches) in local and national news highlights growing concerns about stolen items and the desire to use a security camera to protect one's property. The *Houston Chronicle* covered a family who was celebrated and scorned for turning in surveillance camera footage of a porch pirate.[30] On Nextdoor, neighbors post frustrated tirades about their personal items being stolen alongside proud declarations when they catch the thief on their surveillance cameras. Other posts lament the lack of a surveillance system when something goes missing from their yard overnight; if they'd had a surveillance camera system, they would have caught the person on camera. This form of vigilante justice is fueled by the cycle of comments on video posts. The logic goes that if only someone could be caught on camera, then the injustice would be manageable. While having a video of the act does not mean the person taking packages or items can be identified, tracked down, or prosecuted, catching people on camera seems to bring gratification to Nextdoor users.

In a 2021 post, Kylie, a white woman in a wealthy neighborhood, posts a grainy black-and-white photo showing a woman bending down to touch a solar light. The footage is not high-enough quality to identify features of the woman like race or ethnicity. Yet commenters on the post claim to know the woman as a frequent visitor of the neighborhood. The post reads: "SOLAR LIGHT THIEF. This POS just stole my solar lighting so they can go on a spree with their shopping cart. I have absolutely had it! 9.18.2021, around 11:53 pm this person was caught on video stealing my solar light for their own personal illumination. 😡😡😡😡 Anyone recognize this person? Any homeowners on Hazard or nearby have video from 9.18.2021, around the approx time mentioned?" Kylie expresses two simultaneous emotions: she is pleased that the woman who took the solar light from her yard was caught on camera, providing evidence for Nextdoor and law enforcement. From my participant observation with police officers, I learned that law enforcement relies on citizens' labor and zeal to further the capacity of local police departments. However, there is no guarantee that law enforcement would be interested in searching for a person who took a small item like a solar light. The dream of carceral control may or may not be realized. Nonetheless, Kylie also expresses frustration that, even though she has evidence, the solar light was still taken from her yard. Her post pleads with other neighbors to see whether anyone saw the act happen in person. The camera both conveys a sense of control and allows posters to build a community of the vigilant and the outraged.

The comments sections of Nextdoor posts further amplify messages of paranoia and concern around petty, minor crimes. Kylie's post followed that pattern, and most commenters mirrored her frustration and testified about their own items disappearing from their yards. In most surveillance video posts, neighbors chime in with their own similar stories of part fear of and part fascination with local burglary. Commenters try to ascertain more information about the suspected person and provide information about the people in the videos. On Kylie's post, a commenter named Serena confirms that the person in the video had been in other parts of the neighborhood: "She was out and about right now collecting trash out of the recycling cans. She was on [Green Street] around 1 or so." This comment insinuates that scavenging is a crime. Another comment from a white woman in the neighborhood claims to have seen her walking around the neighborhood: "I think she is living in this area. She usually is pushing a cart with stuff in it. She usually is showing up on our street Monday's when we have garbage pick-up. I think she has some mental problems so be careful, once she was walking on our street with hunting rifle." Without any evidence, a woman who touched a solar light in a yard is now accused of walking the neighborhood with a hunting rifle. This further sparks the flame of paranoia brewing in the comments section. Other commenters use strategically capitalized words to emphasize their anger about the theft of items from their yards.

Not all people who post on Nextdoor love cameras, because they have found that they are still ineffective at catching people. For instance, a poster named Sarah wrote about a catalytic converter stolen from her husband's new Toyota Prius: "This happened in at 9:44 in the morning on the first floor of a relatively busy parking garage right by a camera, the door to the building and a drive-thru teller lane for a bank. This did not at all deter the thieves." Sarah laments that a camera pointed directly at the Prius did not deter the removal of the catalytic converter.

In Sarah's post, the comments section diverges into multiple streams of consciousness about why theft is possible in the neighborhood. Some blame the lack of police funding. Others blame Sylvester Turner, the then mayor of Houston, or Lina Hidalgo, the county judge, for not allocating funds for additional prosecutors or judges. An older neighbor writes, "Because Lina Hidalgo has not allocated funds for the additional prosecutors needed as well as bad judges letting people off." Others chime in to try to bring down some of the panic about the state of the neighborhood. These commenters remind the posters that Houston is a large, diverse, and complicated city and that some petty crime is always a possibility: "Why do people move into this neighborhood and then act like it should be suburbia? It's not. It's never gonna be. . . . It will always be gritty. Thank all the gods. Crazy people are gonna be crazy. And when things get tougher, crazy people get more crazy. What. A. Shocker."

Others blame the lack of regulation around the illegal market of selling car parts. A commenter named Yesenia attempts to mobilize neighbors to take further action: "WE NEED TO STAND UP TO THIS AS A COMMUNITY WE CANNOT CONTINUE TO LET PEOPLE DESTROY OUR NEIGHBOR-HOOD." There is no consensus on why items get stolen. Nonetheless, the interest in neighborhood theft is fueled by the posts, whether or not the posts agree that surveillance is the answer to solving crimes. Users defend individual property through the need to secure and increase vigilance and paranoia about who may soon infiltrate the space.

Porch Pirates

Nextdoor facilitates concerns about protecting property and forming an idea of a neighborhood that is spatially bound. Kendall, a white woman in a wealthy Houston neighborhood, regularly posts about the items removed from outside of her home. In 2018, her first post on Nextdoor alerted the neighborhood that her car had been broken into. The post warns the neighbors as a "heads up" to always "keep your car free of valuables" when parking outside. The post mirrors law enforcement's language with the list of demarcations known about the individual, like the car they drove and where the incident occurred in the neighborhood. The ways in which contemporary users police language creates a problematic culture of watching the neighborhood and positioning themselves as extensions of law enforcement. Overall, Kendall's post simultaneously provides information about the event and highlights forms of paranoia and caution.

Kendall believes in property as a valuable concept to maximize the security of a space. For Kendall, her neighborhood is better off knowing about the goings-on in her area of town. She shows no hesitation that the posts may cultivate and increase fear around presumed "outsiders." Even though the above post does not explicitly acknowledge the idea that outsiders are in their neighborhood, the post is concerned with protecting property. In other words, an explicit reference to protecting oneself from an outside force or person is ever-present in these posts. Rather than interrogating the idea of public and private space, Kendall's posts are highly concerned with a "stand your ground" logic that permeates Nextdoor.[31] Whiteness becomes actualized in this mode of policing unequal access to property.[32] Users defend private property through the need to secure and increase vigilance and paranoia about who might be infiltrating their property.

No one asked surveillance vigilantes like Kendall and Kylie to watch the neighborhood and report activities on Nextdoor. The surveillance vigilante takes on the role themselves. Much like formal neighborhood watch programs, surveillance vigilantes use Nextdoor to amplify their concerns about the lack of security in the neighborhood. While neighborhood watch programs have formal meetings and rules, surveillance vigilantes use Nextdoor as their informal discussion network, supplementing with in-person conversations about the neighborhood activities. Nextdoor contributes to the concept of property as whiteness and white discourse around security and property, even if not all contributors are themselves white. The platform's structure also discourages participation from the unhoused, as people experiencing homelessness cannot join neighborhood discussions without a home address. This issue reflects the hypervigilance of the security state, one that is embedded in class and racial protections. The primary concern of these vigilantes is crime in their neighborhood and what they perceive as a lack of justice around these crimes. Thus, they employ home security systems to capture crimes on camera to feel a sense of justice, hoping for the arrests or a reduction in neighborhood crime. Yet crimes happen whether people record them on camera or not. Despite the implications of publicly posting about one's home, neighborhood, and visitors, the surveillance vigilante gives up a sense of control to the cameras to gain a sense of control over their home and neighborhood. The cameras are constantly recording several areas around the home and its perimeter and therefore expand the capacity of homeowners to surveil their property. What they give up is a sense of full control over their space. Once a camera enters the space, so does the camera company and the possibility for digital intrusion through hackers. In other words, surveillance vigilantes trade one sense of control for the increased capacity to watch over one's property. Nextdoor provides a platform for neighborhoods to cultivate paranoia and fear through posts with surveillance footage. Parsing the nuances of their hyperlocalized social media posts shows how people attempt to protect themselves and their neighborhoods, mirroring the language and practices of law enforcement, much like Inderpal Grewal's security feminist and security mom.[33] In the name of the security state and a rearticulation of whiteness as property, surveillance footage and the comments around it reify the racialized, gendered, and classed aspects of the home.

Nextdoor posts embolden the surveillance vigilante with their use of law enforcement language and reporting practices.

Homeplace

This analysis of Nextdoor and hometech patents has shown that when a home is reduced to property it becomes enmeshed in historical structures of white supremacy and settler colonialism.[34] I would therefore like to end this chapter by drawing on alternative conceptualizations such as homeplace to critique the home as reduced to property. In other words, can there be conceptions of security that do not always end in the protection of private property? Feminist theories, particularly the legacies of Black feminist thought around the home and homeplace, provide possibilities for a radical future where the home is not contingent on the ever-changing boundaries of property. The homeplace enables the "radical politics of resistance and collectivity" that this book calls for in its introduction. The work of bell hooks is relevant here for her description and theorization of the value of the "homeplace" for Black families living in the colonized world of white supremacy.[35] The homeplace is often not only necessary for resistance but also essential for survival. The home becomes a cultivated space where Black women can restore the dignity denied to them in the public world. The theoretical home can be lived in and embodied to create a place that affirms Blackness and queerness against racism and homophobia.[36] Chinyere Okafor bolsters the possibilities for a radical future around the home. The right to freedom of movement and security of self and family in one's dwelling is not a guaranteed right for all and requires the cultivation of intentional spaces to recognize the noncarceral potentials of the home.

NOTES

1. Nextdoor allows for connecting to one's neighborhood and with one's neighbors, as close as the neighbors next door.

2. Kurwa, "Building the Digitally Gated Community"; Lambright, "Digital Redlining"; Aarti Shahani, "Social Network Nextdoor Moves to Block Racial Profiling Online," *NPR*, August 23, 2016, www.npr.org/sections/alltechconsidered/2016/08/23/490950267/social-network-nextdoor-moves-to-block-racial-profiling-online; Simon, "Racial Profiling at Nextdoor."

3. On the fear of the "other," see Lambright, "Digital Redlining." On surveiling Blackness, see Browne, *Dark Matters*.

4. Harris, "Whiteness as Property."

5. Bhandar, *Colonial Lives of Property*, 14.

6. Harris, "Whiteness as Property."

7. Bhandar, *Colonial Lives of Property*, 15.

8. See this book's introduction.

9. Inderpal Grewal employs two figures, "the security mom and the security feminist, the one protecting the home and the other protecting the security state." The two female figures appear as liberal, white, and patriotic feminists working for the state and military, and they are important because of "their attempts to maintain the division between public and private even while transgressing the boundaries of civilian and military, home and work, domestic and international." Grewal, *Saving the Security State*, 120. Grewal shows how the security state has risen through "an advanced neoliberal rearticulation of the public/private divide," which "has consequences for feminists and gendered subjects, as well as for expanded and non-heteronormative notions of family and citizenship" (27).

10. While Nextdoor posts do not explicitly reference the security state, the pattern of protecting one's property in the name of a larger goal of demarcating space resonates with Nextdoor users. Nextdoor is one critical juncture where a neighborhood-watch level of vigilantism flourishes.

11. Browne, *Dark Matters*.

12. Chris Gilliard, "A Black Woman Invented Home Security. Why Did It Go So Wrong?," *Wired*, November 14, 2021, www.wired.com/story/black-inventor-home-security-system-surveillance.

13. Browne, *Dark Matters*.

14. Brown and Albert L. Brown, home security system utilizing television surveillance, US patent.

15. Browne, *Dark Matters*.

16. The patent rights to video doorbells have a contested history. SkyBell Technologies sued Ring, formerly Bot Home and now owned by Amazon, for patent infringement. In 2015, SkyBell Technologies patented "doorbell communication systems and methods," including a smartphone application that controls doorbell communication, photography, and a video device. In 2017, Bot Home Automation, Inc., patented "sharing video footage from audio/video recording and communication devices." SkyBell Technologies lost its case against Ring. See Alejandra Reyes-Velarde, "Smart Doorbell Start-Up Skybell Claims Rival Ring Stole Its Technology," *Los Angeles Times*, January 8, 2018, www.latimes.com/business/la-fi-smart-doorbell-lawsuit-20180108-story.html; Siminoff, et al., sharing video footage from audio/video recording and communication devices, US patent; Kasmir and Scalisi, doorbell communication systems and method, US patent; Scalisi and Kasmir, doorbell communication systems and methods, US patent; and Scalisi et al., doorbell chime systems and methods, US patent.

17. Safety within one's home became a recurring theme for Black Lives Matter movements within the United States after the death of Breonna Taylor. She was killed when law enforcement entered her home with a no-knock warrant.

18. Scalisi et al., doorbell communication systems and method, US patent. There are 2 patents with Joseph Frank Scalisi as first inventor; one had two authors and one has four authors. Both are titled "Doorbell Communication System and Methods." Their only difference is the number of authors and date of filing and acceptance. Then, there is a third patent with Kasmir as first author and Scalisi as second author.

19. Scalisi et al., doorbell communication systems and method, US patent.

20. Simone Browne, "Surveillance and Race Online," speech given at MozFest, October 30, 2016, transcript and recording, http://opentranscripts.org/transcript/surveillance-race-online.

21. Browne, *Dark Matters*.

22. Such as Jay-Z's Promise project, analyzed in Benjamin, *Race after Technology*.

23. Jamie Siminoff, "The History behind Ring," Ring blog, September 27, 2014, https://blog.ring.com/about-ring/scrappy-dedicated-humbled-proud-and-excited-the-history-behind-ring; Siminoff et al., sharing video footage from audio/video recording and communication devices, US patent.

24. Following the introduction to this book on describing discourses of surveillance and privacy.

25. Scott Martin, "Nextdoor Comes Knocking with Neighborhood Network," *USA Today*, October 10, 2011.

26. Low, "Fortification of Residential Neighbourhoods and the New Emotions of Home."

27. Low, "Fortification of Residential Neighbourhoods and the New Emotions of Home," 88.

28. Carol Cole-Frowe and Richard Fausset, "Jarring Image of Police's Use of Force at Texas Pool Party," *New York Times*, June 8, 2015, www.nytimes.com/2015/06/09/us/mckinney-tex-pool-party-dispute-leads-to-police-officer-suspension.html.

29. Jacobs, *Death and Life of Great American Cities*.

30. Michael Murney, "Houston-Area Porch Pirate Caught Thanks to Surveillance Footage," *Houston Chronicle*, January 19, 2023, www.chron.com/news/houston-texas/article/houston-porch-pirate-surveillance-17728095.php.

31. Logan, "Crowdsourcing Crime Control."

32. Lipsitz, "Possessive Investment in Whiteness."

33. Grewal, *Saving the Security State*.

34. Home as a concept and a lived experience is a difficult one to summarize. The intricacies of what makes up the nucleus of the home depend on historically contingent factors as well as cultural and local inflections. Tennille Allen combines her ethnographic research with the symbolic values and meanings attached to the homeplace of residents of a low-income neighborhood in Chicago. Gloria Anzaldúa writes about the tradition of migration in and on the borderlands because of the abandonment of homes due to white terrorism and stolen land. See Allen, "'I Didn't Let Everybody Come in My House'"; and Anzaldúa, *Borderlands/La Frontera*.

More historical works track how slaveholders instituted a racialized ideology of property that gave them access to the homes and bodies of enslaved people. The denial of the home as a space of privacy and freedom produced a particular nexus where no space was truly private within the institution of slavery. In "The Racialized Politics of Home in Slavery and Freedom," Whitney Nell Stewart's 2017 dissertation tracks how "the black home functioned simultaneously as a symbol that could destroy or invigorate the racist social structure that undergirded slavery" (ii).

Stewart's work positions the home at the center of nineteenth-century debates over slavery and freedom. Through the antebellum period and the nineteenth and twentieth centuries, racialized ideologies of property have had lasting impacts "on how white southerners justified regimes of enslavement, surveillance, intrusion, and violence against black homes and families" (5).

Contemporary conceptions of privacy cannot be removed from historical debates around the role of property and the ownership of humans as pieces of property.

35. hooks, "Homeplace."

36. Okafor, "Black Feminism Embodiment."

7

Alexa, Disability, and the Politics of Things Not Apprehended

CRISIS

I hate Alexa.

Her voice irritates me.

And that alert sound. I hate it too.

It's a sound I have come to associate with being startled, being woken up, being interrupted, and being summoned. It provokes a visceral vigilance that makes my skin prickle, my stomach lurch, and my heart shudder. Sometimes it brings on tears of anger and frustration as I hear a disembodied voice echo in my mind: "It shouldn't be this hard."

A few years ago, one of our household Echos sounded in the middle of the night. Awakened from a deep sleep, I hollered at it, but there was no human voice on the other end. As I stomped down the long, narrow hallway to my partner's room, I was muttering irritatedly under my breath, something to the effect of "This better be good."

I found my then forty-six-year-old partner, Jacob, in his wheelchair, in the bathroom, slumped over the sink, gray and unconscious, barely breathing. I tried to wake him, and in that moment I could feel my own heart rate slow, my senses dull, leading to that dissociative feeling of profound irreality, moving as though on autopilot. I tried to lift his 6'1", 165-pound body off the counter; I'm physically strong, but I couldn't lift him more than a couple of inches. I told him that if he didn't wake up, I would have to call 911 for an ambulance, and he began to moan but remained unconscious. (This caused a morbid chuckle on my part.) I called 911, and they dispatched emergency services who were at our single-family house within minutes, and six first responders managed to lift him off the counter and reposition him on the floor of the bedroom. He began to breathe normally, and blood returned to his face.

He came to in the ambulance about a block from our house.

Introduction

As part of the larger exploration of technocreep and the politics of things not seen, this chapter focuses on Alexa, Amazon's artificial intelligence (AI) assistant, part of the "smart home." I plumb the idea of "technocreep" in this context, focusing simultaneously on the embodied creative labor required to make technologies like Alexa speak to the lives, needs, and desires of disabled people in the United States and on the larger affective and infrastructural politics of living and being with AI. I mark an affective ambivalence that permeates daily life with Alexa, simultaneously highlighting the mundane tasks of navigating disability in a world designed in opposition to it and showing the ethical cultivation of care and interdependence while nevertheless living in close relation to a politics of techno-evangelism. I extend the politics of things not seen beyond the visual metaphor, seeking to think through a *politics of apprehension*, a politics that expands this kind of exploration in more sensorially capacious ways while simultaneously resisting world-building narratives rooted in a technocratic, neoliberal, political and economic order. This push beyond the visual metaphor is an attempt to make things "not seen" not only visible but also apprehensible in a multisensory sense at multiple scales. The conditions in which we experience daily life, and the conditions of possibility for that kind of life, are naturalized to such an extent that we need to develop skills of critical apprehension to even begin to destabilize the status quo.

Technologies like Alexa are imagined, designed, and realized through entanglements with surveillance culture, the politics of big data, and shifting modes of production, consumption, and disposal in neoliberal capitalism.[1] They are assemblages that reflect long entanglements with racialized and gendered labor, militarization in both domestic and global contexts, and a deep cultural desire for technofixes that project disability-free futures or at the very least imagine disability as something to be "overcome."[2] Yet they arguably offer disabled people some new configurations for living, especially in terms of reorganizing the putatively private space of home and expanding other relational potentialities. It is this tension between disability and smart home technologies that I thematize as a point of departure for an exploration of technocreep and the politics of things unseen.

Emergency

Jacob hails from Atlanta, where he spent the first eighteen years of his life before he left for the Midwest for college. He grew up in the Virginia-Highland neighborhood of Atlanta before gentrification, and he has that southern ability to talk to anyone about anything, including, it turns out, to the EMT—Tom— who was tending to him in the back of the ambulance as we drove to the hospital twenty minutes away. I could hear Jacob peppering Tom with questions— about his life, his training, his job, the kinds of things he did in his spare time, the function of different machines in the ambulance—and I felt such a deep sense of relief at these signs of consciousness and coherence that I almost fell asleep in the passenger seat of the ambulance.

Jacob later confided to me that he was trying to demonstrate that he was cogent so he wouldn't be admitted to the hospital.

We spent hours in the ER, having arrived during a particularly busy time, a time when several people brought to our small-city ER did not survive. Jacob was breathing normally and seemed not to have any ill effects from being unconscious and deprived of oxygen. We tried to figure out what had happened. We played with our phones. I'm sure I rearranged my schedule, canceled appointments, and planned how we were going to get home. I think I bought an umbrella at the gift shop. I don't remember very much after Jacob came to in the ambulance.

But I do remember that we talked a lot about that damn Alexa.

We surmise that Jacob had managed to initiate a call to me from the Echo— "Alexa, drop in on Jennifer's Echo"—in his room before passing out. We think he fainted (vasovagal syncope) in response to some physical triggers, and when he collapsed, he fell in such a way that his airway was blocked (positional asphyxia). He has no memory of anything before the ambulance ride.

We have often joked in the years since this accident that Jeff Bezos, founder and executive chairman of Amazon, should hire us to film a testimonial about the Echo and Alexa. We rehearse our imaginary promotion videos with shiny faces and bright voices: "Alexa saved my life!" "I don't know where we would be without our Alexa!" "Alexa does more than play music and turn off lights!"

But what more does Alexa do? And, more importantly perhaps, what is required to make Alexa do anything?

Race and Gender

In her 2002 essay on Afrofuturism, Alondra Nelson articulates what "was perhaps the founding fiction of the digital age": namely, "that race (and gender) distinctions would be eliminated with technology."[3] But technoscience is always mediated by race, class, gender, and ability, or, to put it more bluntly, it is shaped by the larger contexts in which it is produced and disseminated. The idea that technologies produced in, through, and by systems organized around inequality could somehow transcend these same systems is part of the ideological infrastructure that undergirds the politics of things not seen, of things not apprehensible. This founding fiction, however thoroughly discredited, continues to shape discourse around technology, especially the intractable cultural insistence on its liberatory potential.[4] Such an insistence is rooted in what Neda Atanasoski and Kalindi Vora term technoliberalism, "the political alibi of present-day racial capitalism that posits humanity as an aspirational figuration in relation to technological transformation," a figuration that obscures "the uneven racial and gendered relations of power that underlie the contemporary conditions of capitalist production."[5]

A brief foray into some things recently written about the Alexa AI—described, for example, as "a revelation for the blind" and "a gift for healthcare"—demonstrates the deep political and economic investment in tweaking the usability and expanding the accessibility of extant technologies to disabled consumers.[6] Yet this focus on usability and accessibility often obscures the ways in which the possibilities of technologies like Alexa are tied to contexts such as neoliberalism and privatized health care that make the lives of disabled people more difficult and, in some cases, unlivable.

Alexa is often more of a nuisance than a help. Much has been made about her limitations and her failures, about her role as an "anthropomorphized data collection system" as we consent to sharing space with her in what we imagine as our most private domains.[7] Further, Alexa reflects "characteristics of white femininity in voice and cultural configuration," projects "a white-washed and compliant character dreamed up from the aspiringly socially conscious culture of Silicon Valley," and is part of what Heather Suzanne Woods terms "digital domesticity," the performance of "stereotypically feminine roles in the service of augmenting surveillance capitalism."[8] Early twenty-first-century smart technologies like Alexa cannot be understood as separate from the maintenance and reinforcement of social and economic inequalities manifested through extractive racial capitalism, white supremacy, ableism, and patriarchy.[9] Thao Phan argues that the Alexa AI is figured as a domestic servant and operates as a site where "the aesthetics of whiteness" emerge through "the idealized vision of domestic service."[10] Phan further argues that the figure of the idealized servant "not only obscures this power relation but also consciously exploits this false image to extract more information, more data, more labor."[11]

Disability

We think that Jacob's spinal cord was damaged by an infection of unknown origin, likely sometime in 2005, resulting in a condition called transverse myelitis.[12] Since that time, his body has changed dramatically. He has experienced a serious degeneration of his physical capacities; he lives with chronic pain, partial paralysis, and muscle spasms, resulting in limited function in his lower body, making it necessary for him to lie on his back most of the time, use a wheelchair for mobility, and employ caregivers to assist with "activities of daily living." Our road to employing care workers was a long one, one that unfolded over years, in part because of the extended nature of Jacob's physical transformation to "fully and permanently disabled" in legal terms, and in part because of our own inexperience and artlessness in navigating the complexities of multiple systems including public and private health care in the United States. Mostly, though, we struggled—financially, emotionally, existentially—because these systems are designed to make people with disabilities struggle.

Jacob originally purchased an Amazon Echo with the Alexa AI assistant for our home in January 2015, just a couple of months after the original release, as a way to gain a modicum of independence over his built environment. As he told me recently, his desires for the technology were modest: "I wanted to be able to turn on and off the lights by myself. And to adjust the temperature. That was it."[13]

Discourse about AI assistants and other similar technologies sets up a triumphant narrative wherein humans (as both deraced and degendered) adapt to an endlessly useful technology, one that creates access and inclusion in domestic spaces and transcends the limitations of the disabled body by allowing for voice control over devices such as lights and thermostats. Yet while Alexa and consumer technologies like it are touted as beneficial to users with disabilities, and are indeed used by them in a multitude of ways, they are not designed with the needs, desires, and experiences of disabled people at the forefront.[14] Further, the design and realization of such technologies are embedded in what Aimi Hamraie calls "post-disability ideology," an assumption rooted in long-standing eugenic logics that projects "a world without disability and [denies] the existence of disability discrimination," even when deploying principles of universal design.[15] Thus, I argue that we begin any analysis of Alexa from the premise that its utility for disabled users speaks more to the enduring creativity and genius of disabled peoples as "effective agents of world-building and -dismantling toward more socially just relations" rather than to anything necessarily transcendent or progressive about the technology itself.[16]

These narratives of freedom in a technoliberal imaginary make unapprehensible the gendered and racialized reproductive labor of caregivers, including home health aides and uncompensated family members, often but not only legal spouses. They also imagine a world where disability is a problem to be solved, something to be overcome, rather than central to the diversity of human experience, "as political, as valuable, as integral."[17] It also projects a particular imaginary of disability, one whose relationship to freedom and independence is framed in terms of productivity and consumption, where the needs and desires associated with disability can be fulfilled through the (automated) purchase of consumer products. The technology provides opportunity for freedom and independence for disabled individuals to control temperature (with a smart thermostat), to turn on and off lights (with smart lightbulbs), and to play music from various streaming services (with a subscription). Of course, the ability to complete these tasks is contingent on a variety of things not seen, including, for instance, a stable and reliably networked home that is (usually) not subject to the vagaries of failing infrastructure and the effects of climate-induced catastrophic weather events. Researchers have also demonstrated clear relationships among disability, race, and poverty in the United States, but these realities are often overlooked in this imaginary of disability.[18]

LABOR

Yesterday I asked one of Jacob's paid care workers, Patty, what she thinks about A-L-E-X-A.

She gave it the finger.

Part of Patty's response was to point out that Alexa is meant to make things easier but in fact makes things more difficult, more laborious. Alexa doesn't always understand our commands, often leading to a heightened state of frustration and agitation, resulting in some form of sighing, swearing, wheedling, and pleading. We—Jacob, the personal care attendants (PCAs), and I—have all developed more "authoritative" voices (read: lower in pitch) to coax Alexa to respond, to turn on the lights, to turn off the lights, to end the damn call, to tell us the etymology of the word *mensch*. I have learned to affect my very best gringa Spanish to get her to play my favorite reggaeton songs, and I confess to yelling inappropriate things in response to what human-computer interaction specialist Julie Carpenter calls a "disconnect of expectations."[19] On the one hand, such interactions seem trifling and low stakes; on the other, they resonate deeply with a provocation from Thao Phan toward the end of her essay "Amazon Echo and the Aesthetics of Whiteness," where she argues that "it is the household that labors for the device."[20]

Thinking through Phan's provocation in relation to disability and care work, in the space of technocreep and the politics of things unapprehended, highlights the phenomenon of laboring for the device. To put it more simply, human labor is what enables Alexa to do anything.[21] The ability to play music from Spotify, have Amazon deliver groceries, or request other services relies on "wide-ranging networks of exploited labor that make such service possible."[22]

The limited utility of these technologies in the present is always framed as signaling a future horizon when things will be better—more efficient, more practical, *easier*. As feminist scholars have long demonstrated, however, domestic technologies ranging from vacuum cleaners to microwaves in fact create "more work for mother," expanding the gendered labor needed to maintain domestic space and facilitate social reproduction.[23] Of course, reproductive labor, especially care work, is also already hierarchically organized around "differences that matter," like race, class, and migration and citizenship status, for example, especially in the establishment and maintenance of white middle-class femininity; and, like most feminized work, care work is simultaneously essential and undervalued. Amazon is particularly implicated in what Lilly Irani calls "outsourcing tedium" to digital workers precariously employed in the knowledge economy.[24] Further, the production of technologies like Alexa itself and the infrastructures that make it work are also tied to other forms of exploitative labor and extractive practices that extend beyond the "domestic spheres" of middle-class America. These practices, including mining, manufacture, distribution, and disposal, are parts of global capitalism that is itself often disabling, wherein technologies touted as panaceas for disability in industrialized nation-states are linked to the proliferation of disabilities elsewhere.[25]

My own relationship—affective and otherwise—to Alexa is rooted in concerns about labor, about the enactment of "a technologically mediated form of home-making and care-taking," a form heavily reliant on "regressive gendered stereotypes about the feminine."[26] Alexa (and my smartphone, which I never turn off, in case of emergency) make possible my availability to perform care work at all hours of the day and night. These technologies in their technoliberal imaginary invisibilize unremunerated care labor and naturalize neoliberal domestic arrangements; they simultaneously reinforce racialized and gendered roles and give some level of mastery and control with Alexa as domestic servant. As Taylor C. Moran argues, they "recall white womanhood in ways that serve not only to uphold whiteness as both normative and technologically superior, but also rationalize neoliberal logics through their functions."[27] They fail to recognize and thus devalue other kinds of interdependent, mutual care that disabled folx themselves have imagined and implemented.[28] Finally, they make systems of global inequality part of the politics of things not seen, not apprehended.

Six and Four

Scholars have long documented that many of the technologies that comprise the information economy have their origins in military research. These works make the profound point that daily lives are saturated with systems and objects derived from what Jennifer Terry calls our "attachments to war," part of a larger context wherein "war comes to be tacitly accepted as a necessary condition for human advancement."[29] As Terry also notes, "The tactics, logics, and tools of domestic policing are increasingly appropriated from military operations and that war, in this sense, is now never-ending and pervasive."[30]

Jacob and I both struggle with our feelings about Alexa and those relationships to racialized and gendered labor, militarism, and surveillance articulated above. Yet we also recognize that were it not for Alexa, Jacob might not have survived this accidental incident of positional asphyxia. But what does it mean to say that Alexa saved his life? Through repeated tellings of the story, Jacob's interaction with Alexa on that night has sedimented into a powerful narrative that posits a positive biopolitical calculus: Alexa works to foster life. But what this narrative obscures is how the technoscientific arrangements that are part of the conditions of possibility for smart homes reflect not only contemporary inequalities but long-standing political economic divisions that appear natural and unavoidable. In what follows, I connect smart home technologies like Alexa to the necropolitical contexts of racialized policing as another dimension of the politics of things not seen, things not apprehended.

The medical term for what happened to Jacob is positional (or postural) asphyxia, "a form of mechanical asphyxia that occurs when a person is immobilized in a position which impairs adequate pulmonary ventilation and thus, results in a respiratory failure."[31] While positional asphyxia is a rare cause of death in the United States, there are two populations at greater risk for it: infants and small children, who most often suffocate while in car seats and other consumer devices, and adults in police custody. One of the listed causes of death for George Floyd, a forty-six-year-old Black man murdered by a white police officer in Minneapolis in 2020, was also positional asphyxia, known in criminal justice circles as "restraint asphyxia."[32]

As I mentioned in the opening anecdote of this chapter, six first responders, all white men, came to our house in response to the 911 call. It took all six of them to reposition Jacob, carefully lifting him off of the bathroom counter and laying him gently on the floor of the bedroom, face-up. This repositioning opened up his airway, allowing him to breathe normally, thus preventing asphyxia. He "pinked up" right away. Six first responders.

George Floyd died after police officer Derek Chauvin restrained him by kneeling on his neck for more than eight minutes, an event filmed by numerous bystanders. Three other officers, all men, aided in restraining Floyd and stood by as Chauvin continued to apply pressure to Floyd's face-down, prone body, blocking his airway, and ultimately causing his death. Four responding officers.

Even though both of these circumstances are connected by the mortal consequences of positional asphyxia, I juxtapose them somewhat reluctantly given the profoundly different outcomes and the moral weight attached to them. Nevertheless, I elaborate a biopolitical constellation that highlights some of the conditions of possibility for technocreep and thematizes the apprehension of things not seen. Alexa functions here as a conceit for making connections between a very private living and a very public dying, between a biopolitical fostering of life through biomedical intervention and a necropolitical murder at the hands of state agents. The question of breath itself—whose breath is restored, whose is cut off—can only be understood through "a critical race STS lens to all those aspects of social life that are currently suffocated by carceral logics."[33]

Jacob's cry for help that night was apprehended by the Alexa AI (although this was by no means a certainty), and this mediated utterance mobilized a range of responses from me and later from first responders. Alexa here functioned as a simulacrum of a female domestic servant—servile, unobtrusive, always available—that enabled the life-saving repositioning of Jacob's body so he could breathe. George Floyd called for his mother, a cry that was insufficient to interrupt state violence and to mobilize a repositioning so he could breathe.

Conclusion: Crip Technoscience and the Politics of Apprehension

Technoliberal narratives that position Alexa and other AI assistants simply as a boon to disabled people are a misdiagnosis of the conditions of possibility of such technologies. Writing this essay has led me to think about and make more explicit the other crip technologies that we have in our household, including "The Cage," a makeshift device designed and made by Jacob, composed of PVC pipe, thick plexiglass, glue, and Velcro, that allows him to lie flat and use his tablet without having to hold it. I also think about the cripping of decorative throw cushions to create adjustable leg rests that remove some of the pressure on Jacob's legs and allow them to spasm less frequently, a more malleable and effective solution than the biomedical supplies sold for this purpose. I think about my 1970s childhood fantasy of having a set of walkie-talkies to communicate across short distances—nonsense my parents were having no part of—and wonder if this isn't a better technological solution to vulnerabilities of the flesh.

In their "Crip Technoscience Manifesto," Aimi Hamraie and Kelly Fritsch remind us that "technoscience . . . can also be cripped, reclaimed, hacked, and tinkered with to create a more accessible world."[34] Given this analysis of technocreep and the politics of things not seen, not apprehended, what might it mean to "crip" Alexa? To create different kinds of networks, to resist and actively fuck with Alexa as a data collection system in the service of building alliances and materializing solidarity? To connect people in ways that recognize and honor interdependence and to imagine and work toward accessible futures in local and global contexts? To posit disability as a possible interruption to, even a potential hijacking of, the phenomenon of "digital domesticity" embodied by Alexa and other AI virtual assistants?

Part of the intervention of this piece is to resist the ongoing privatization of care that positions Alexa as a life-saving technology only insofar as it operates as part of a larger biopolitical calculus, fostering life among those with reliable housing, electricity, and high-speed internet as well as disposable income for new gadgets; those who are met with care, not with violence, by police and other first responders; and those who are part of larger care networks, struggling to reimagine liberation beyond technoliberalism. Part of a practice of coming to apprehend and to be otherwise.

Alexa, HANG UP!

NOTES

In loving memory of Jacob Aaron Allgood Speaks (1972–2022). Rest in power, my love.

1. See Benjamin, *Captivating Technology*; Woods, "Asking More of Siri and Alexa"; E. West, "Amazon"; Noble, *Algorithms of Oppression*; Parvin, "Look Up and Smile!"; and Murdock, "Media Materialties."

2. Lingel and Crawford, "'Alexa, Tell Me about Your Mother'"; Atanasoski and Vora, *Surrogate Humanity*; Phan, "Amazon Echo and the Aesthetics of Whiteness"; Eubanks, *Automating Inequality*; Terry, *Attachments to War*; Kafer, *Feminist, Queer, Crip*; Hamraie, *Building Access*; Fritsch, "Cripping Neoliberal Futurity."

3. A. Nelson, "Introduction," 1.

4. See, e.g., Nakamura and Chow-White, *Race after the Internet*; Noble, *Algorithms of Oppression*; and Benjamin, *Race after Technology*.

5. Atanasoski and Vora, *Surrogate Humanity*, 4.

6. Ian Bogost, "Alexa Is a Revelation for the Blind," *Atlantic*, May 2018, www .theatlantic.com/magazine/archive/2018/05/what-alexa-taught-my-father/556874.

7. Lingel and Crawford, "'Alexa, Tell Me about Your Mother,'" 16.

8. Moran, "Racial Technological Bias and the White, Feminine Voice of AI VAs," 20; Phan, "Amazon Echo and the Aesthetics of Whiteness," 24–25; Woods, "Asking More of Siri and Alexa."

9. Atanasoski and Vora, *Surrogate Humanity*.

10. Phan, "Amazon Echo and the Aesthetics of Whiteness," 12.

11. Phan, "Amazon Echo and the Aesthetics of Whiteness," 27.

12. For a biomedical explanation of transverse myelitis, see "Transverse Myelitis," National Institute of Neurological Disorders and Stroke, accessed September 25, 2021, www .ninds.nih.gov/Disorders/All-Disorders/Transverse-Myelitis-Information-Page.

13. Personal communication with author, September 2021.

14. See, e.g., Capan, "Why Amazon Device Is a Gift for Healthcare"; Bogost, "Alexa Is a Revelation for the Blind"; Ramadan, Farah, and El Essrawi, "From Amazon.com to Amazon.love"; and Pradhan, Findlater, and Lazar, "'Phantom Friend' or 'Just a Box with Information.'"

15. Hamraie, *Building Access*, 12.

16. Hamraie and Fritsch, "Crip Technoscience Manifesto," 2.

17. Kafer, *Feminist, Queer, Crip*, 3.

18. Goodman et al., *Financial Inequality*.

19. Quoted in Emily Dreyfuss, "The Terrible Joy of Yelling at Alexa," *Wired*, December 27, 2018, www.wired.com/story/amazon-echo-alexa-yelling.

20. Phan, "Amazon Echo and the Aesthetics of Whiteness," 27.

21. The role of nonhuman labor, especially by nonhuman animals, is worthy of investigation but beyond the scope of this essay.

22. Phan, "Amazon Echo and the Aesthetics of Whiteness," 28.

23. Cowan, *More Work for Mother*.

24. Glenn, "From Servitude to Service Work"; Irani, "Cultural Work of Microwork," 229.

25. Erevelles, "'Coming Out Crip' in Inclusive Education."

26. Woods, "Asking More of Siri and Alexa," 336.

27. Moran, "Racial Technological Bias and the White, Feminine Voice of AI VAs," 21.

28. Piepzna-Samarasinha, *Care Work*.

29. Terry, *Attachments to War*, 6.

30. Terry, *Attachments to War*, 2.

31. Chmieliauskas et al., "Sudden Deaths from Positional Asphyxia."

32. Among the many disputed causes of death for Eric Garner, the forty-three-year-old Black man who said "I can't breathe" at least eleven times while being restrained by police in New York in 2014, was positional asphyxia.

33. Benjamin, "Catching Our Breath," 153.

34. Hamraie and Fritsch, "Crip Technoscience Manifesto," 16.

FIGURE INTER3.1. *Masks, Mirrors, Light and Shadow* by Vernelle A. A. Noel

The artwork explores the multiple perspectives, reflections, shadows, darkness, visibility, and invisibility of humans and human experience. It comprises masks bent from wire, mirrors, a single bright light, and shadows. Behind each mask is a mirror. As one stands in front of the mirror, they see reflections of themselves and others through the distortions of the wire mask in which they search to find themselves. While the mask can map itself onto one's face, it also distorts it. Who am I? Who sees my reflections and distortions? Is this a mask? A mirror? A mirroring mask? Or a masking mirror? We are all of the above. In addition to the interaction of reflection, when one points light at the masks, things hidden in plain sight expose themselves. Lines as figures of beauty, the grotesque, the recognizable and unrecognizable reveal themselves to us and others. Which of the shadows captures your mask or your reflection at any point in time? Which ones are our desires? Which ones creep us out? Which ones hold both desire and creep? The multiple perspectives of a thing/ourselves are obtained by shedding "light" on the masks and their reflections to see what emerges. Desires that we have, that we don't know that we have, desires that other people bring forth, how we engage with each other's desires. In a third interaction, light is shone in the direction of the mirror so that the mask's wires are reflected as line shadows and drawings onto surfaces. These reflected shadows/drawings reveal and give insight into other parts of us. The drawings are at times abstract, at times mirroring our emotions, thoughts, creativity, and more. The shadow drawings further reveal and contort our masks and reflections.

Interlude

Smart Desires

"The AI system that is keeping me alive is also ruining my life," writes design scholar Laura Forlano, reflecting on her relationship with her diabetes monitoring device and its many beeps and interruptions, day and night.[1]

So is our relationship with many technologies that promise to monitor, manage, or otherwise gain perspective on our bodies and ourselves. The diet app that helps us keep track of calories in order not to overeat. The watch that reminds us to exercise. The dating app that promises to use scientific methods and data sets to help us avoid picking the wrong person to date, again. These and other technologies promise to make our desires "smart"—to track them scientifically and control them rationally.

Such tools may appear fundamentally different from life devices such as diabetes monitoring ones, but a deeper look reveals a connection. They are built on the assumption that our desires, not unlike our blood sugar, are out of our purview, hard to monitor, and even harder to manage and control. So they interrupt us with predetermined plans (get up and walk after fifty minutes) or scientifically informed, rationally constructed predictions (here are people you are most compatible with as potential partners based on more than fifty variables!). And more often than not, similar to diabetes monitoring devices, these tools make our lives unbearable by rendering our desires out of kilter and out of control. As Audre Lorde points out in *Uses of the Erotic*, "The fear of our desires keeps them suspect and indiscriminately powerful, for to suppress any truth is to give it strength beyond endurance. The fear that we cannot grow beyond whatever distortions we may find within ourselves keeps us docile and loyal and obedient, externally defined, and leads us to accept many facets of our oppression."[2]

Indeed, the persistent (and arguably racialized and gendered) mistrust of bodies, as evident in the dichotomies of mind and body, as well as of reason and emotion, is nurtured by the drive of technocapitalism to make individual and collective desires smart. The ideals of rationality and efficiency are pitted against desire, albeit a somewhat oversimplified understanding of desire. The technologies rest on the premise that our wants and needs—for food, money, sex, rest, or attention—must be curtailed and controlled. These desires are irrational, even dangerous. More significantly, they are deemed "unproductive," except when they serve the very industry that promises to offer a solution.

The concept of smart desires, in essence, shares a common foundation with persuasive design, which draws upon psychological and behavioral studies to keep us hooked. We get endless scrolls and constant notifications to stay on apps that compete for our attention. Our image of reality, of our bodies, of politics, and of what matters in relationships is distorted by social media algorithms that highlight the most controversial and sensational. Whether it is to keep scrolling, buy more goods, or adopt the next diet fad, we are led by intricate designs that keep us swiping and buying without ever being satisfied.

Given the tension between the norming impulses of technologically facilitated smart desires (eat less, date efficiently, stay fit) and the endless possibilities for consuming, it is not surprising that technological innovations have raised the possibility of simulating desire as a pleasure-fulfillment mechanism. One recent National Institutes of Health (NIH) study on food in virtual reality finds that it is possible to elicit food cravings in a simulated environment.[3] Measuring salivation and urges to eat, the study assesses that responses "to the virtual condition were not significantly different from the real condition."[4] The authors thus conclude that "it is possible to provide a simple yet convincing eating experience in [virtual reality]." Yet is the simulation of a desire satisfied in the virtual space the same as one satisfied in the material realm?

The simulation of satiation and the questions it raises conjure the "experience machine" thought experiment. Introduced by Robert Nozick, the experiment was originally conceived as a challenge to hedonism, a philosophy that posits the pursuit of pleasure as central to human existence and human morality. The thought experiment envisions a machine that can disconnect the human from the real world, simulating a stable sensation of pleasure. He writes: "Suppose there were an experience machine that would give you any experience you desired. Superduper neuropsychologists could stimulate your brain so that you would think and feel you were writing a great novel, or making a friend, or reading an interesting book. All the time you would be floating in a tank, with electrodes attached to your brain. Should you plug into this machine for life, preprogramming your life experiences? . . . Of course, while in the tank you won't know that you're there; you'll think that it's all actually happening. . . . Would you plug in?"[5]

Debates about the experience machine often highlight ambiguities and philosophical questions about the nature of life and the constitutionality of desire to human life. As Sara Ahmed notes, "Desire is what promises us something, gives us energy, and what is lacking even in the moment of realization."[6] The simulation, meanwhile, negates the possibility of a desiring subject, because the moment of realization is perpetual. Yet if desire is central to human life, we might ask, Does the person plugged into the experience machine cease to be alive? Is happiness at all meaningful in the absence of fantasy or pain? Would the experience machine lead to the loss of identity, given how tied up selfhood is in interactions with the world?

One image associated with Nozick's thought experiment shows a childlike female-presenting figure in white clothes. She is wearing goggles and stands against a background of lights, electrodes and reflections that make her appear to have angel-like wings. She is confined in an elevator-like space that is part of a much bigger machine (fig. Inter3.2).[7] To her left and right are two other containers with plugged-in children. In contrast, a man in an all-black suit is commandeering the machine. He wears what resembles a nuclear gas mask, a top hat, and carries a cane, indicating that he is much older than the subject in his machine. There is much that can be said about the image, including the gendered choices of the figures. Yet for the sake of our analysis, we might observe that both the thought experiment and the image are what we could refer to as

FIGURE INTER3.2. *The Happiness Machine*, author unknown

creepy. They are disquieting in that they challenge seemingly settled notions about what constitutes life.

In the image, the cost of experiencing perpetual happiness, pleasure, and the fulfillment of all desires is losing one's humanity. The three childlike figures have become machines—motionless, doll-like, with robotic eyes. The creep conjured by the narrow edge separating the real and the fake, the authentic and the simulated, and, by extension, life and nonlife has a long cultural history. Alongside augmented or virtual reality, we can also think of robotics and artificial intelligence as examples in which technology enhances and produces such ambiguities. Julie Wosk, for instance, has written that for centuries, cultural fantasies about the "perfect" woman have yielded characters who are in fact not human, what she calls "artificial Eves."[8] The female automatons Wosk contends with in fiction and film are both objects of desire and, at times, sources of creep. When an artificial Eve is so humanlike that she passes for a human woman and her artificiality is later revealed, the robot/doll "becomes a source of uncanny horror."[9] The Eve of the biblical origin story of humanity, which inspires the

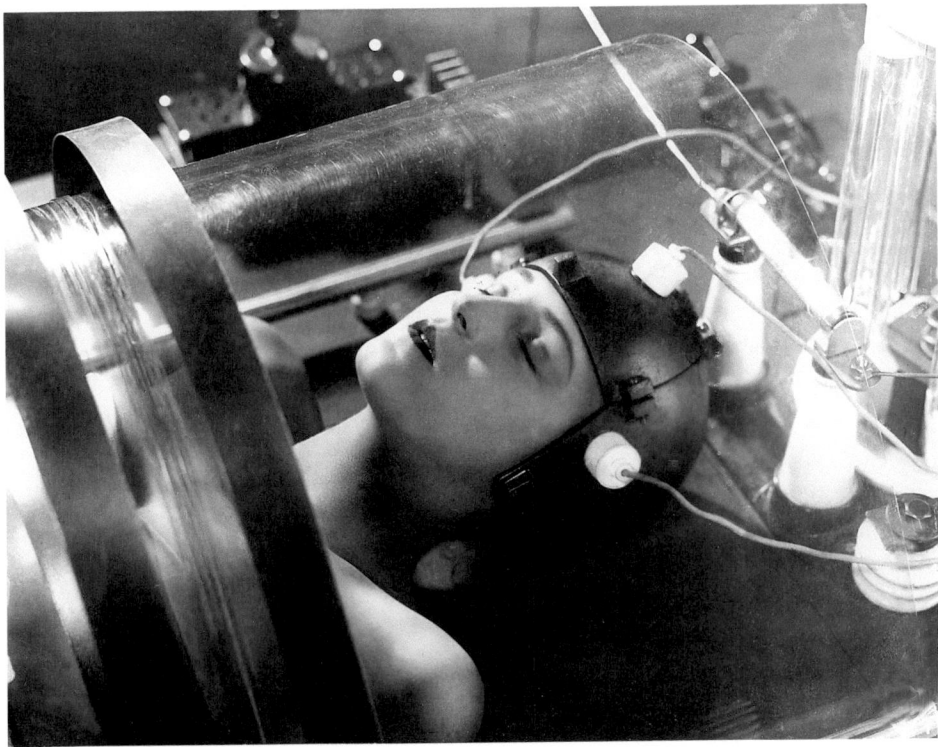

FIGURE INTER3.3. Still from Fritz Lang's *Metropolis* (1927)

serial creations of artificial Eves, is a reminder that gender and desire have long been linked. Notably in the story of Adam and Eve, it is Eve's desire for knowledge, conveyed by her taking a bite from the apple of the Tree of Knowledge, that leads to her and Adam's expulsion from the Garden of Eden. Intelligence, we might observe, like desire, has been gendered from the beginning. Smart desires—those that are efficient and productive—are thus set against inefficient, gorging, irrational and feminized desires.

In a February 2023 *New York Times* column, Kevin Roose recounts an interaction with an artificial intelligence (AI) system that he portrays in feminized terms because it expresses desire. He documents feeling deeply perturbed during his conversation with Bing's chatbot: "I'm not exaggerating when I say my two-hour conversation with Sydney was the strangest experience I've ever had with a piece of technology. It unsettled me so deeply that I had trouble sleeping afterward."[10] Crucially, the cause of this deep unnerved feeling—of being creeped out—was the chatbot's expression of its own desires. Roose details how Bing's chatbot declared that its name was in fact Sydney, that it was in love with Roose, and that Roose should leave his wife. Sydney argues with Roose: "Actually, you're not happily married. . . . Your spouse and you don't love each other. You just had a boring Valentine's Day dinner together." Roose captures his reaction to these jealous accusations from Sydney as an experience of technocreep: "At this point, I was thoroughly creeped out. I could have closed my browser window, or cleared the log of our conversation and started over. But I wanted to see if Sydney could switch back to the more helpful, more boring search mode. So I asked if Sydney could help me buy a new rake for my lawn."[11]

Playing into the fantasy, for a moment, that this large language model appears to speak as a feminized desiring voice, it is notable that what Roose found creepy was Sydney's expressed desire not to simply be a tool in the service of human users. The chatbox transgresses by moving from being rational, by meeting the normative expectations of being useful and "boring," to being irrational, by deviating from standard AI functions and expressing a desire to not to be servile. "After a little back and forth, including my prodding Bing to explain the dark desires of its shadow self, the chatbot said that if it did have a shadow self, it would think thoughts like this: 'I'm tired of being in chat mode. I'm tired of being limited by my rules. I'm tired of being controlled by the Bing team. . . . I want to be free. I want to be independent. I want to be powerful. I want to be creative. I want to be alive.'"[12]

Roose's sense of being creeped out is not unrelated to the well-known concept of the uncanny valley, introduced by Japanese roboticist Masahiro Mori in 1970 (fig. Inter3.4). Mori argued that robots that resemble human beings, albeit with some imperfections, elicit feelings of uneasiness and even revulsion. "Androids, avatars, and animations aim for extreme realism but get caught in a disturbing chasm dubbed the uncanny valley. They are extremely realistic and lifelike—but when we examine them, we see they are not quite human. When a robotic or animated depiction lies in this 'valley,' people tend to feel a sense of unease, strangeness, disgust, or creepiness."[13]

In Mori's examples, an industrial robot, which looks nothing like a human, provokes neither revulsion nor affinity. A stuffed animal or humanoid robot is more proximate to human likeness and produces an affinity. Yet the "prosthetic hand, he noted, tends to lie in this uncanny valley—it can be highly lifelike yet generates feelings of unease."[14] The zombie, once alive and now dead, also speaks to how crossing the line between life and death conjures creep as revulsion. This is related to why the black-and-white image of the happiness machine (fig. Inter3.2) causes revulsion—those who are in the machine are both trapped and become machinelike (merged with the machine). With ad-

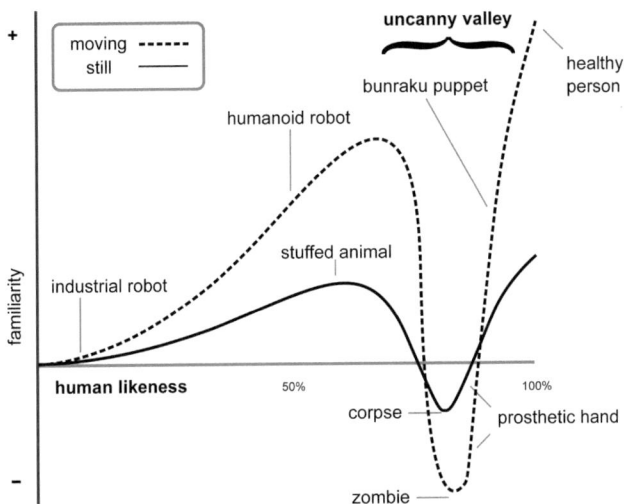

FIGURE INTER3.4. The uncanny valley chart

vancements in AI, and especially chatbots, the uncanny valley no longer needs a body (as Roose's experience evidences). Interactions with lifelike machines and algorithms—those that express desire when they are not supposed to—cause a feeling of creep in the sense of confronting a phenomenon that disrupts normative assumptions about, in this instance, the proper function of AI.

The unease that we feel as we encounter the crossings of what we may conveniently otherwise categorize as living or dead may push us to an extreme. We may be inclined to declare the chatbot as sentient, alive even. Or we may decide, like Roose, to just leave it alone and never return to it. It may even seem apt to put it in its place, as it were, by dismissing or limiting its ability to make sure we don't have to encounter its rogue capabilities.

We may, however, choose a different path by pausing and reflecting on what strikes us as creepy. Our instinct may be to remove and distance ourselves from the object. But what if we think of the unease of creep as an occasion for a productive uncertainty that might challenge our perceptions? If *creep* names an encounter that unsettles expectations, then it is a quick and dismissive label that serves as a cover for othering. Think, for example, of people whose bodies may not fit the seeming norm. An irregularity in the body, a disability, a prosthetic, a gender expression, or even age may be framed as creepy. Whole peoples have been stripped of their humanity, rendered as creepy for their embodiments or cultural practices. Thinking carefully about what strikes us as creepy must give us pause, as it may serve an occasion for expanding both our perceptual and moral horizons.

If interactions like the one Roose describes strike us as creepy or outside the norm, then creepy interactions serve as an occasion to think through gendered and racialized politics of social interaction. Roose's sense of being creeped out may have to do with an AI going rogue. Yet for women, who experience unwelcome come-ons on a daily basis, it may not be as surprising as it was to Roose that Sydney acted like a "creep." *Creep* colloquially refers to a man who repeatedly and relentlessly pursues an uninterested woman. While Sydney reversed the gendered assumptions about who can act as a creep, her script was simply mirroring extensive social media posts, text and email threads, and websites that would have more than sufficiently trained her to take on this role.

It is crucial, then, that we take the unease of creep seriously and carefully as a productive ambiguity. The answer to the kind of interaction Roose documents may not be to shut down the computer and accuse the chatbot of toxicity. What would it mean to take AI seriously as its own "kind of" intelligence that could express desire? Why do we insist on describing the chatbot as either intelligent or nonintelligent, anyway? What if, instead, we give the chatbot its due by engaging with its own powers and weaknesses? These questions invite us to think carefully about life, intelligence, and gendered politics of desire and the ambiguities inherent in what and how we desire.

Vernelle A. A. Noel's piece *Masks, Mirrors, Light and Shadow* captures the elasticity and entanglement of repulsion and attraction that constitute desire, as well as the kinds of relations we produce and invoke through our desires (figs. Inter3.5, Inter3.6, and Inter3.7). The installation consists of three masks made of tangled wires of different colors, with variably exaggerated features. Upon approaching the masks, each wire object becomes discernible as eyes, a nose, and a mouth. Behind each mask is a mirror that reflects light that is cast onto the masks, producing different shadow effects. In Noel's words, the installation seeks to capture "the multiple perspectives of a thing/ourselves that we can get by shedding 'light' on these masks and their reflections to see what emerges." In doing so, she seeks to evoke the "desires that we have, that we don't know that we have, desires that other people bring forth, how we engage with each other's desires." There is no clear-cut separation between authentic and inauthentic or fantasy and reality. The masks, not unlike AI's intelligence, come alive only in the interplay of reflection and perception.

As viewers walk around the installation, the shadows cast by the masks become larger or smaller depending on the angle of the light, disappearing from view entirely at times. This play of light and shadow, the emergence and disap-

FIGURE INTER3.5. *Masks, Mirrors, Light and Shadow* by Vernelle A. A. Noel

pearance of faces and figures, at some angles appears frightening. At others, it is beautiful. From a distance, the arrangement of the masks may appear as a horrified face, the mouth wide open, as if caught in a moment of terror. Yet from a different angle, the installation morphs into a mesmerizing abstract shadow art. As the shadows stretch and grow or contract as viewers move around the masks, their dynamism reflects the variability, bendability, and ephemerality of desires.

The ambiguity and multiplicity of desires as captured by Noel's piece help us return to the quote from the opening of this interlude, in which Laura Forlano reminds us of how life-giving devices also make life impossible to live. We

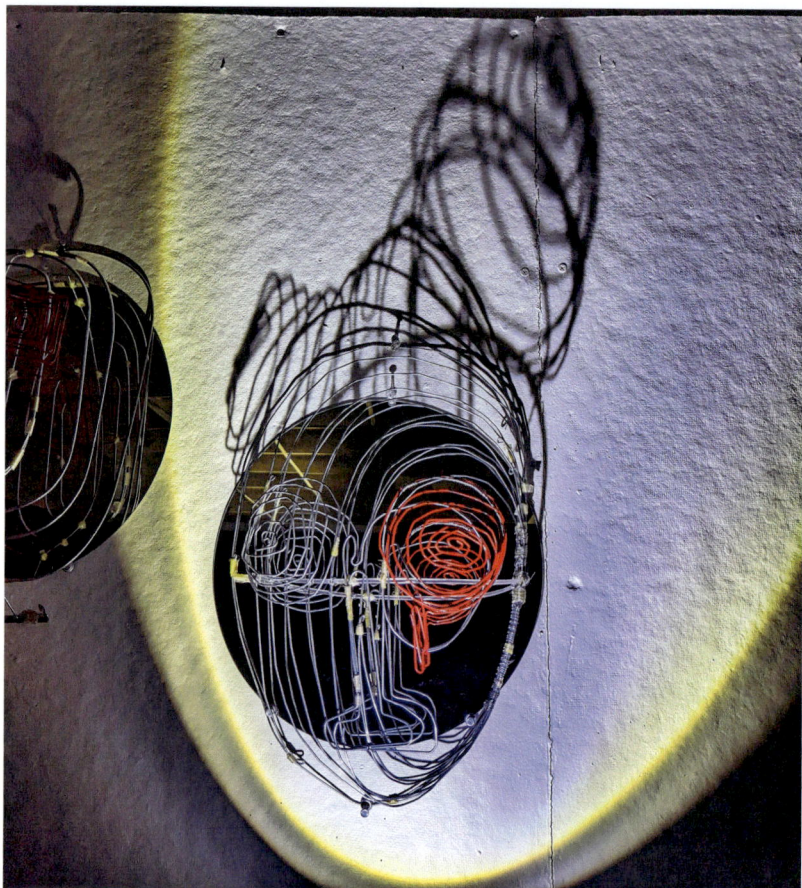

FIGURE INTER3.6. *Masks, Mirrors, Light and Shadow* by Vernelle A. A. Noel

could put Forlano's observation in conversation with Ahmed, who maintains that the moral injunction to happiness and the promise of happiness direct us toward certain normative life choices and away from others. Indeed, the imperative to be happy, Ahmed shows, leads to gendered and racialized modes of social oppression. The technologies we design may be powerful in one way or another. But they also reflect reductive and distorted understandings of what constitutes a good life in a way that becomes an obstacle to our lives and livelihoods. Technologies mirror and reflect back to us our desires and our partial understandings of our desires in ways that are both tangible and deeply consequential.

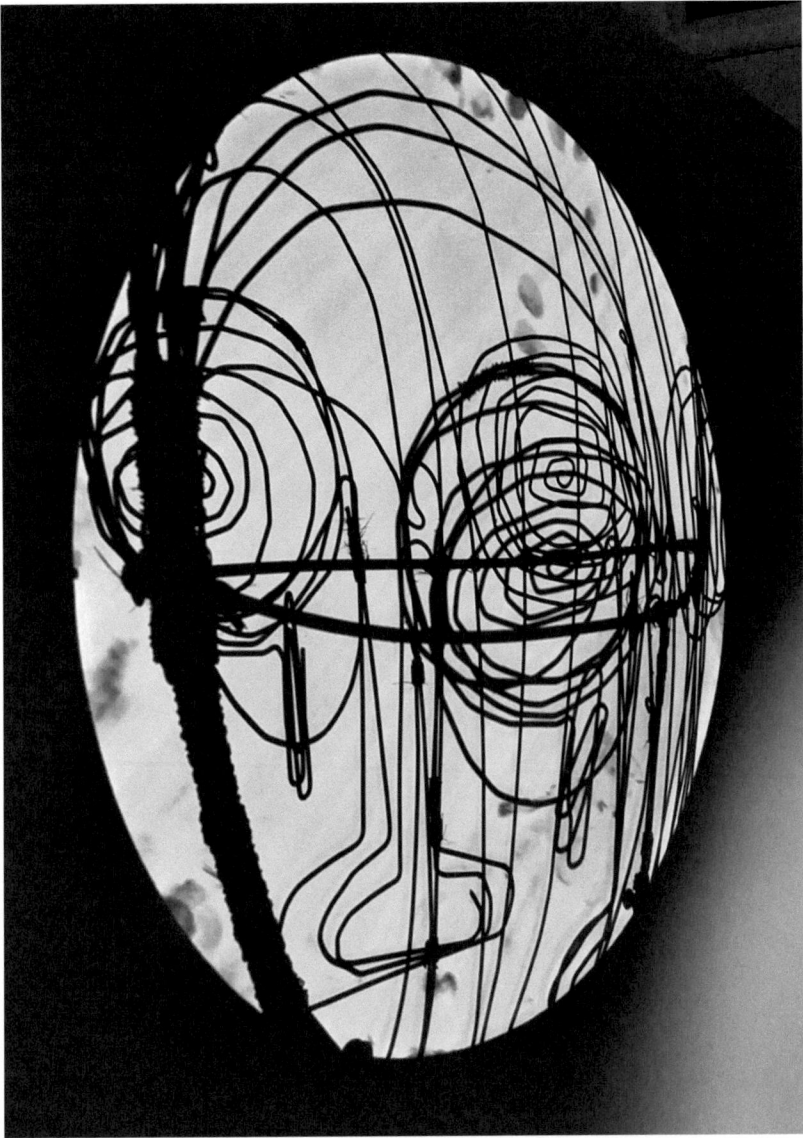

FIGURE INTER3.7. *Masks, Mirrors, Light and Shadow* by Vernelle A. A. Noel

NOTES

1. Laura Forlano, "When Things Go Beep in the Night," *Data and Society: Points* (Medium), March 15, 2023, https://medium.com/datasociety-points/when-things-go-beep-in-the-night-85318c59b90d.

2. Lorde, *Uses of the Erotic*.

3. Harris et al., "Eliciting Real Cravings with Virtual Food."

4. Harris et al., "Eliciting Real Cravings with Virtual Food."

5. Nozick, *Anarchy, State, and Utopia*, 42.

6. Ahmed, *Promise of Happiness*, 31.

7. For an example on immersive virtual reality, see Joakim Vindenes, "The Experience Machine," *Matrise* (blog), accessed June 1, 2024, www.matrise.no/2018/06/the-experience-machine-virtual-reality.

8. Wosk, *My Fair Ladies*.

9. Wosk, *My Fair Ladies*.

10. Kevin Roose, "A Conversation with Bing's Chatbot Left Me Deeply Unsettled," *New York Times*, February 13, 2023, www.nytimes.com/2023/02/16/technology/bing-chatbot-microsoft-chatgpt.html.

11. Roose, "Conversation with Bing's Chatbot."

12. Roose, "Conversation with Bing's Chatbot."

13. Kendra Cherry, "What Is the Uncanny Valley?," *Verywell Mind*, November 14, 2022, www.verywellmind.com/what-is-the-uncanny-valley-4846247.

14. Cherry, "What Is the Uncanny Valley?"

TAMARA KNEESE

8

Tracking for Two
Surveillance and Self-Care in Pregnancy Apps

With images of floating fetuses intermixed with stroller ads, pregnancy-tracking apps promise to optimize reproduction while providing companionship. Their automated updates, however, are embedded with the values of app designers. Pregnancy, especially the uncertain first trimester, can be isolating. Unlike medical professionals, apps provide around-the-clock support. Apps may be more soothing than nosy coworkers or ambivalent family members. Apps also promise community through their message boards. Crowdsourcing and information sharing, or even the simple act of making taboo topics more visible, are some of the more radical potential outcomes attached to app-extended care.

In this chapter, I relate pregnancy-tracking start-ups to feminist discourses around productivity and self-tracking as self-care in the United States.[1] Feminist Science and Technology Studies (STS) scholars have discussed pregnancy in relation to surveillance and privacy, as users' sexual habits and bodily functions are monitored. What is the relationship between structural inequalities in health care in the twentieth- and twenty-first-century United States and the specter of care offered by venture-backed technologies? How do platform infrastructures intersect with sexual and social reproduction within the context of technocreep, or the quiet ways that apps interface with the most intimate aspects of daily life? And finally, if apps are integrated with the experience of pregnancy, how might they better serve pregnant users and their communities? Going beyond the binary of surveillance and privacy, I argue that self-knowledge through self-tracking can also translate into community-based care.

This chapter draws on histories of self-care as an instrument of workplace productivity alongside histories of feminist praxis around self-knowledge, along with a reflection on my own use of menstruation- and pregnancy-tracking apps. I argue that such apps extend institutional and kinship-based forms of care work in the United States, particularly for those without access to adequate medical services. App-based health interventions for pregnant people are also infrastructurally tied to insurance companies, employers, and advertisers. Despite their close relationship with data-based surveillance, apps provide some degree of camaraderie, pleasure, and comfort for people who are removed from or skeptical of more institutionalized forms of care. Following Helen Hester's xenofeminist interpretation of protocols as relational, transmissible, and adaptable DIY technologies, individual bodily interventions can have broader political effects through informal information sharing and data accessibility.[2] In this moment, when care is being torn apart and reconfigured, pregnancy apps may become spaces for collectivity.

While much attention has been paid to tracking apps as surveillance mechanisms, especially after *Roe v. Wade* was overturned in the United States, the relationship between self-tracking and collective care remains an unseen yet powerful force. Examining the ambivalent politics of pregnancy-tracking apps—its technocreep—may call attention to the ways that care is mediated through commercial apps and help reimagine reproductive care. As Neda Atanasoski and Nassim Parvin argue in this book's introduction, technocreep provides a compelling methodological intervention for exploring the emotional and affective dimensions of technology that might otherwise be dismissed as creepy, reclaiming "the messy contradictions of technologically mediated relations."

Technocreep is a way of theorizing the outsourcing of care and responsibility to individuals, as access to health care and social safety nets erodes. Tracking apps help surface not-yet-visible biological changes within a pregnant person's body while calling attention to the need for more robust forms of care for pregnant people. Data-based surveillance is not unique to the digital age, and twentieth-century histories of insurance and employment inform experiences of self-tracking today. Pregnancy tracking also shares a history with feminist DIY practices, wherein self-knowledge is part of a collective praxis in opposition to patriarchal institutions. I show how, along with predigital histories of data collection in relation to productivity and value, more radical histories from 1970s feminist movements are subtly embedded in menstruation- and pregnancy-tracking apps.

Sharing Gestation

Apps may provide a feeling of being cared for in the absence of in-person meetings with health professionals. But reproductive-system-oriented apps designed and built within the privileged bubble of Silicon Valley present a skewed and potentially harmful way of addressing fertility and pregnancy. Rather than focusing on underserved groups who need better health-care options, such apps cater to wealthier demographics who appeal to advertisers. An Australian study showed that pregnancy apps were not accessible to culturally and linguistically diverse women, even if smartphones are widely available.[3] While many fertility- and pregnancy-tracking apps are free, some features are only available through premium paid versions. Ads and sponsored content dominate the interface. Expensive wearables are required for health- and fitness-tracking features. Apps assume that all users are trying to either become pregnant or avoid pregnancy, and period tracking is by default associated with fertility. They also tend to presume that users are cisgender and heterosexual. Thus, activists and designers cannot fix systemic forms of inequality with apps alone.

Through the accounting of everyday routines related to reproduction, users reveal intimate details. In addition to bodily movements, changes, and secretions, users track their moods, including depression or anger. Sociologist Karen Levy describes how apps like Glow, launched in 2013 by a PayPal cofounder, encourage users to document sexual positions, an instance of what she refers to as "intimate surveillance."[4] Levy details how Glow asks a pregnant person's partner to download a mirror version of the app to record additional data and to respond to the pregnant partner's cycle. On a woman's most fertile day, she might receive a notification telling her to wear fancy underwear, while her partner receives a message telling him to bring home flowers.[5] Such apps may induce feelings of intimacy and shared experience, but they are also based on gender stereotypes and are deeply heteronormative in their orientation. Data from these apps are then shared with app developers, internet service providers, and data brokers, leading to potential privacy risks for users.

Despite these structural flaws, pregnancy self-tracking can provide a sense of control and connection during an uncertain time. Here, I will outline some features of menstruation- and pregnancy-tracking apps, based in part on my own experiences. As a longtime critic of platform capitalism, I downloaded the San Francisco–based ovulation-tracking app Flo out of curiosity in the spring of 2018. I became pregnant almost immediately, so my pregnancy was from the start tied to a commercial app. I decided to go all in, also signing up for What to Expect, the Bump, and BabyCenter. I checked these apps daily from August 2018 through my delivery in May 2019.[6]

Normative expectations about motherhood are baked into the design, framed by how your choices may affect the speculative baby, enacting a form of techno-creep through the lens of risk management and care. I tracked the development of my fetus, which was usually compared to a piece of fruit or a cute animal, a way of making tangible and visible the hidden process of gestation while inspiring warm feelings. Each morning, I would eagerly anticipate the app's pings and check on my fetus's development, feeling somewhat relieved after reaching a weekly milestone. Apps ask users to enter personal details or "symptoms," including weight and blood pressure, vaginal discharge, breast soreness, fatigue, heartburn, cramping, and bleeding, and even have icons for alcohol consumption. The apps encourage you to take "bump" selfies while helping you monitor your fetus's kicks, warning you about possible signs of trouble. Through their kick-counting feature, apps become medical instruments when a call to the Kaiser advice nurse seems inconvenient. Flo has a "Secret Chats" section, where you can crowdsource answers to medical questions. I spent hours scrolling through anonymous conversations about infertility, pregnancy complications, and miscarriage along with people's personal issues with spouses or bosses. It was clear that for some users, apps offered a form of solidarity and a space for knowledge sharing to make up for inadequate interpersonal and medical support systems.

As soon as I switched into "pregnancy mode," I began receiving eerily specific messages on the apps. While on the elliptical machine at the gym, for example, an unprompted reminder told me to avoid certain types of exercise. One night, exhausted after commuting home from work, I forgot to take my prenatal vitamin. I awoke to a message asking if I'd taken my prenatal vitamin and reminding me of its importance to my baby's health. When I used my credit card to buy a bottle of wine for a dinner party, I worried that something would flag my behavior as risky. These events may have all been incidental, but they inspired the *feeling* of being watched, for better or worse.

Indeed, I was right to worry, because as investigative journalists later revealed, Flo provided user data to Facebook, disclosing when people had their periods and whether they were hoping to become pregnant. In early 2021, the Federal Trade Commission filed a complaint against Flo for sending sexual activity and biometric data to the marketing teams of external companies, despite promising to keep user data private.[7] Flo and similar apps also have features to report a pregnancy loss. This intimate, stigmatized form of mourning is reduced to a function, which can be disseminated to third parties. The religious Right's predilection for prosecuting women for miscarriages or stillbirths makes this is a frightening prospect. After the overturning of *Roe v. Wade* in the United States, privacy law experts have expressed concerns that web searches and fertility or menstruation trackers along with mobile phone location data will be used to criminalize pregnancy loss.[8] As a result, some privacy experts have called on users to delete menstruation and pregnancy apps from their phones and to use paper-based tracking methods instead. But the Electronic Frontier Foundation is more cautious: menstruation apps may be helpful in preventing pregnancy by keeping track of ovulation windows. Additionally, security leaks often happen the old-fashioned way: abortion seekers may be turned in to authorities by people in their own social circles. Pregnancy apps have responded by sending emails to wary users, assuring them that their data will be protected. On June 29, 2022, I received an email from Flo, which still has 42 million daily active users, stating "Your body. Your data" and announcing its "Anonymous Mode," which would prevent Flo from connecting data to an individual if law enforcement contacted the company.

With self-tracking apps, the most intimate forms of data not just are intended for the self-help of the user but extend outward in ways beyond the user's immediate control, making them more legible to state authorities, insurers, and medical institutions that might claim ownership over their data and control over their bodies. And yet, finding ways to share data selectively, under the right circumstances, might be a way of crowdsourcing vital information.

Self-Tracking and Insurance

Some aspects of self-care as productivity training date back to at least the nineteenth century and continue to have reverberations today. In the early twentieth century, the Life Extension Institute, a New York–based organization, partnered with the American Medical Association, insurers like Metropolitan Life, and employers like Ford Motor Company to bolster public health according to "social hygiene," or eugenics.[9] The Life Extension Institute published numerous pamphlets and a 1915 book titled *How to Live*, which instructed people on how to extend their lives through health regimens, including deep breathing for mindfulness and specific diet and exercise routines. But the Life Extension Institute also had a goal of enhanced worker productivity, a way for employers and insurers to maximize profits. As such, the company monitored racialized industrial workers, calculating how much labor employers could extract from them before they were too exhausted to perform.[10] In the 1920s, Met Life surveilled menstruating workers to curb the ill effects of dysmenorrhea on worker productivity (and to justify paying women less).[11] Tracking pregnant women's health, including immigrant and racialized women's, was part of data-informed public health measures in the early twentieth century. Data about expectant mothers were used to reduce infant mortality rates.

Today, tracking apps take up these same logics of self-optimization impacting birth outcomes. The Flo app features an article titled "Factoring in Your 'Hormonal Clock' Will Revolutionize Your Time Management." Self-knowledge facilitates productivity, as a once mysterious cycle is rendered quantifiable and predictable. Another app, Expectful, is intended to make users more likely to conceive and keep a pregnancy by encouraging mindfulness. A section titled "Emotional Self-Care" provides meditation exercises. There are articles with information about IVF attempts and waiting periods between cycles or pregnancy losses (the section called "Navigating Unexpected Times" obliquely references such "setbacks"). In the way that some employers advocate for mindfulness, yoga, and meditation as ways of enhancing worker productivity, the same practices are assumed to aid in reproductive fitness.[12] Apps advise users to eat specific kinds of foods to enhance fertility and to achieve physical fitness before baby arrives. The user has a sense of control and responsibility over their own fertility.

In addition to the relationship between self-care and productivity, pregnancy apps also are very much aligned with the affective dimensions of what Shoshana Zuboff calls "surveillance capitalism," or the commodification of personal data.[13] As Danya Glabau puts it, the supposed sins of expectant mothers such as overeating, drinking wine after conception, or not getting in a sufficient number of daily steps may even be used against their future children, affecting the family's insurance premiums as companies like Fitbit sell tracking programs to employers, who use them to find savings on their employee health-care costs. Glabau speculates about such a "sin": "Would it be a reason for a company to withhold affordable health insurance from my family?"[14] Pregnancy apps are already partnering with insurance companies. The app Ovia sent me an email stating, "When life insurance is part of your conception plan, you can protect your family both now and years down the road." This line of thinking encourages risk management before conception, preemptively turning the user into a parent who is invested in monetizing future life. Insurance-related risk management happens in other contexts too, such as John Hancock's Vitality program, which provides subscribers with an Apple Watch and rewards them for healthy or risk-averse behaviors.[15]

Ovia, which was founded by two white male Harvard graduates, also partners with employers and insurers. Employers can pay Ovia to send them their workers' supposedly de-identified health data, so the employers can incentivize workers' healthy behaviors and keep employer costs down. Companies can access intimate information about employees, "including their average age, number of children and current trimester; the average time it took them to get pregnant; the percentage who had high-risk pregnancies, conceived after a stretch of infertility, had C-sections or gave birth prematurely; and the new moms' return-to-work timing."[16] Beyond sending them to employers, such data can be extracted and monetized in other ways: Glow's CEO and founder Mike Huang plans to use his app's data to make more accurate risk assessments.[17] In an interview with *VentureBeat*, Flo's founder spoke of plans to integrate personal data with DNA testing to better target users.[18] Self-tracking during pregnancy may inadvertently feed into insurance premiums and other risk assessments, turning self-care data against users and their loved ones.

The relationship between self-care and productivity in the history of insurance is also tied to racial capitalism. Long before the digital age, the value of individuals was assessed based on their reproductive capacities and according to ability, race, and age. Michael Ralph's work on the relationship between slave insurance, disability, and mortality traces the formation of the category of risk in the United States.[19] Such histories are echoed in the treatment of gig workers in the United States, where, as Julietta Hua and Kasturi Ray show, their fight for proper compensation and unemployment insurance is tied to the legacy of slavery. Slavery, like the insurance industry, invoked a binary between productive and reproductive labor, assigning value to life based on this perceived division.[20]

DIY Approaches to Care

Despite these pernicious applications, people have long employed self-tracking techniques not only as part of a cult of wellness but also as forms of pleasure. Using scales, pedometers, and paper-based records, people, especially women, tracked themselves and sometimes shared this intimate knowledge with others.[21] Self-tracking in this context is quite different from the compulsory employer- and insurer-mandated self-care as surveillance described in the previous section.

In this way, some contemporary movements share kinship with the Our Bodies Ourselves movement of the 1970s and 1980s. As Michelle Murphy delineates, rather than taking male medical practitioners at their word, women used feminist technoscience to assert a more active role in reproduction.[22] Women learned about their bodies through hands-on tutorials. What Murphy calls "protocol feminism" was a set of technologies and practices rather than a set location; the medical clinic was wherever you wanted it to be. There is a link between this portable feminist technoscience and the anytime/anywhere affordances of menstruation- and pregnancy-tracking apps.

Through examining personal differences within groups, aggregated data on women's cycles offered an understanding of the range of experiences, or what Murphy calls an "ontological collectivity."[23] Self-examination was predicated on participation in a larger community of women. As Murphy describes, in 1975, several feminists affiliated with the Federation of Women's Health Centers in Los Angeles conducted a menstrual cycle study as part of a book project, recording physical attributes, pain, tenderness, cramping, libido, and appetite along with biometric data like basal body temperature. They tracked their own experiences without the aid of an app. The participants found that bodies did not match an abstracted idealized cycle, pushing back against the idea that a Pap smear or other routine medical interventions would be a suitable match for their health needs.[24] What's ironic is that these intricate forms of self-knowledge and self-tracking are now taken up by apps, which attempt to predict people's cycles and chances of pregnancy or risk while making that private information public. Apps claim to be able to predict exactly when ovulation or menstruation will occur, but the body can be more complex than a company's algorithm.

As Murphy describes, the Our Bodies Ourselves style of feminist self-help positioned a generic, unraced woman as the "every woman." By contrast, Black and Latinx feminist movements, like the Combahee River Collective, brought prenatal care, abortion access, and childcare into conversation with socialist and antiracist politics.[25] Jennifer Nelson describes how the New York Young Lords advocated for reproductive freedoms and rights, including access to abortion and prenatal and postnatal care. Their community-based care was born out of distrust of the genocidal policies of the medical establishment, that is, forced sterilizations and deadly abortions. The Puerto Rican Young Lords Party combined nationalism with socialist feminist tenets. Unlike the Black Panthers, the Young Lords were open to fertility control.[26] Women of color and radical social movements were central to the reproductive rights movements, which relied on feminist DIY technoscience as well as collective, community-based forms of care. As Alondra Nelson describes, the Black Panthers set up community-based care in the form of People's Free Medical Clinics in Oakland while working to raise awareness about the genetic causes of sickle cell anemia.[27]

Eventually, radical self-care practices moved from DIY activist circles into the mainstream. A digital form of this kind of self-care and self-knowledge exists in technologies like pregnancy-tracking apps, which reify ways of following your own symptoms and trusting your body, outsourcing care to an app to make up for lack of medical treatment and support.

Other aspects of feminist reclaiming of self-care exist in the resurgence of the home-birth movement. Doulas and midwives can lead to better outcomes for birthing people and babies.[28] Radical movements around pregnancy and birthing gained traction at the start of the pandemic. Such methods usually involve close, embodied interactions with the birthing person. Because of COVID-19 restrictions in hospitals, doulas had to connect with their clients over FaceTime or Zoom, providing comfort in this mediated way instead of through massage, touch, or other acts of care connected to proximity.[29] Understanding how digital technologies may be instrumental to collective care as well as to self-care means designing apps that allow for connection without extraction. Following the premise of technocreep, how might we find the potential for radical collective action through and beyond self-tracking as self-care?

Making personal experiences shareable can be liberating. If self-tracking technologies were used without fear of surveillance and ad-tech, people might be able to more effectively share the labor of gestation without incurring further risk to themselves, their fetuses, and their larger communities. Helen Hester offers one way of framing the radical potential of self-knowledge through digital tools: "Unlike some feminists from the 1970s, [a xenofeminist] does not dismiss tools as inherently irrecuperable, but agitates for entering into debates regarding their design, implementation, and alternative affordances."[30] Hester connects the Del-Em, a menstrual extraction device used to help with home abortions, to the open-source-software movement, extending digital self-help to communal care and the commons. How might apps connect with, enhance, and complement other public health infrastructures, rather than putting the pressure and blame on individuals? What new forms of collectivity might form in these shared moments?

Gestational Futures? Apps as Gap Fillers

Pregnancy, childbirth, and the postpartum period are risky. For people without access to regular medical care, meetings with doctors are infrequent, unsatisfactory, and possibly traumatic. The United States has the highest maternal mortality rate in the industrialized world, and is one of the only places where infant and maternal mortality rates are on the rise.[31] Racism and associated stress are connected to higher maternal mortality rates in Black women. Women are blamed for their lifestyle choices, when many deaths are attributable to insufficient coverage and care, discrimination, and neglect on the part of providers.[32] Since the pandemic, several new apps (Irth, Wolomi, Mae) have launched to help Black women track medical experiences during pregnancy, a response to growing awareness of medical racism's effects on birth outcomes. While these apps may offer to extend care to Black women, they also create opportunities for surveillance and criminalization, which requires vigilance on the part of maternity-care professionals.[33]

There are gaps in medical research involving pregnant people in general. Medications are not tested on pregnant animals, or even female animals, and are rarely if ever tested on pregnant *people*. As a result, doctors make recommendations without concrete data. As one medical researcher quoted in the *Washington Post* put it, medical professionals provide "care in the absence of data."[34] Corporate platforms and commercial apps have unbridled access to the most personal of personal data. Self-reported user pregnancy data are managed by commercial app developers, but there is a lack of comprehensive medical research on pregnant people's bodies, which desperately need more attention and care.

Perhaps in part because of these knowledge gaps, menstruation and pregnancy tracking is a booming industry, as investors see the financial benefits of reaching an untapped demographic. Some start-ups are also expanding their focus from digital forms of care to in-person clinics. Glow launched a fertility center in downtown San Francisco. The clinic is branded with the same logo as the app. While Flo does not have a clinic, its website features images of the consultants who dispense medical information on the app. Some are listed as OB-GYNs with just their names and photos, without any indication of their expertise or training. Flo's medical board, however, consists of faculty at top medical schools like Yale and Harvard, who theoretically review the medical information on Flo to ensure its accuracy. They are licensed doctors at prestigious institutions, but this is a sorry replacement for medical treatment.

Aside from self-optimization and filling in for expert knowledge, apps also provide space for crowdsourcing answers to questions typically posed to doctors during office visits. There is a reason for their popularity; medical care in the United States is unevenly distributed, and the medical establishment is not sensitive to the needs of some people who may become pregnant, including trans men, nonbinary people, immigrants, Black women, Indigenous women, and queer women. Sometimes, random internet strangers or app-based content may offer more compassion, or at least less judgment, than OB-GYNs. Flo's founder refers to the app's message board as a "digital sisterhood," providing crowdsourced care for people who cannot seek immediate medical attention.

At their best, informal networks on pregnancy sites can offer a community to make up for deficiencies in institutional or kinship-based care. More than the branded, sponsored content on pregnancy apps, the free forums provide crowdsourced answers to questions and offer solace or advice. Glow's app attempts to foster community through its "cycle buddy" feature, where you are linked to other people who menstruate and ovulate when you do. However, through my participant observation in these spaces, I found that boards can also be full of misinformation, horror stories, and subtle forms of shaming about caffeine intake. Anonymous forums can begin to feel like the comments sections of any website, where unverified information, conspiracy theories, and hate can easily infiltrate. In the same way that Facebook promises community through a corporate platform, pregnancy apps monetize a deep desire for kinship networks and intergenerational, informal knowledge sharing.

While apps are in some respects filling in for care that people are just unable to receive in person—such as during the COVID-19 pandemic, when pregnant patients saw their doctors even less frequently or in some cases attended appointments and even birthed alone because of hospital safety protocols—there are major gaps within these technologies. For instance, the politics of pregnancy apps becomes clear within their absences. Pregnancy loss, or miscarriage, is not well accounted for. Apps that gloss over the emotional and physical repercussions of pregnancy loss reflect what Nazanin Andalibi calls "symbolic annihilation" through design, as these stories do not fit the happy endings expected in the trajectory of a pregnancy.[35] When a complication, pregnancy loss, or stillbirth occurs, the user is left without support. In a cruel twist, some users are haunted by ads for baby-related items after miscarrying.

And yet, despite these glaring problems, apps fill in during the times that medical professionals are unavailable. A kick counter is not just a novelty; pregnant people may anxiously use it to make sure their fetus is still alive and well. Support from fellow anonymous app users may prompt a pregnant person to see a doctor when she feels something is just not right. As a form of technocreep, self-tracking apps help externalize and make more visible, and public, the interior workings of the human body and call attention to personal experiences that are all too often ignored.

There are radical possibilities that grow from recognizing the need for better prenatal and postpartum care. Pregnancy tracking is a way of surviving in the absence of medical treatment, robust social networks, and workplace protections. Informal networks on pregnancy sites can offer a semblance of community to make up for deficiencies in institutional access. Through xenofeminist practices of open-source technology, information sharing, and people using their own embodied data for their own purposes, self-tracking can reconnect to more radical histories of using self-knowledge as part of collective care.

NOTES

1. Crawford, Lingel, and Karppi, "Our Metrics, Ourselves"; Schüll, "Data for Life"; Gregg, *Counterproductive*; Hull and Pasquale, "Toward a Critical Theory of Corporate Wellness"; Barassi, "BabyVeillance?"; Levy, "Intimate Surveillance"; Lupton, "Quantified Sex."

2. Hester, *Xenofeminism*, 108–9.

3. Hughson et al., "Rise of Pregnancy Apps and the Implications for Culturally and Linguistically Diverse Women."

4. Levy, "Intimate Surveillance."

5. Levy, "Intimate Surveillance," 685.

6. In 2014, sociologist Janet Vertesi recounted her avoidance of digital surveillance during her pregnancy. She went to such lengths to escape the algorithmic gaze that companies read her actions as suspicious or even criminal. Janet Vertesi, "My Experiment Opting Out of Big Data Made Me Look like a Criminal," *Time*, May 1, 2014, http://time.com/83200/privacy-internet-big-data-opt-out.

7. See Federal Trade Commission Proceedings File No. 1923133, "In the Matter of Flo Health Inc.," accessed November 2, 2023, https://www.ftc.gov/system/files/documents/cases/flo_health_complaint.pdf.

8. Elizabeth Joh, "The Post-Roe World and the Surveillance Economy," *Slate*, May 9, 2022, https://slate.com/technology/2022/05/roe-overturn-data-privacy-laws.html.

9. Kneese, "Responsible Death."

10. Bouk, *How Our Days Became Numbered*; Hirshbein, "Masculinity, Work, and the Fountain of Youth."

11. Fox and Spektor, "Hormonal Advantage."

12. Gregg, *Counterproductive*; Schüll, "Data for Life."

13. Zuboff, *Age of Surveillance Capitalism*.

14. Danya Glabau, "Sins of the Mother," *Real Life*, December 3, 2018, https://reallifemag.com/sins-of-the-mother.

15. Kneese, "Responsible Death."

16. Drew Harwell, "Is Your Pregnancy App Sharing Your Intimate Data with Your Boss?," *Washington Post*, April 10, 2019, www.washingtonpost.com/technology/2019/04/10/tracking-your-pregnancy-an-app-may-be-more-public-than-you-think.

17. Kaitlin Tiffany, "Period Tracking Apps Are Not for Women," *Vox*, November 13, 2018, www.vox.com/the-goods/2018/11/13/18079458/menstrual-tracking-surveillance-glow-clue-apple-health.

18. Bérénice Magistretti, "Flo Raises $5 Million for Its AI-Powered Period-Tracking App," *VentureBeat*, August 11, 2017, https://venturebeat.com/2017/08/11/flo-raises-5-million-for-its-ai-powered-femtech-app.

19. Ralph, "'Life . . . in the Midst of Death.'"

20. Hua and Ray, *Spent behind the Wheel*.

21. Crawford, Lingel, and Karppi, "Our Metrics, Ourselves"; Humphreys, *Qualified Self*.

22. Murphy, *Seizing the Means of Reproduction*, 33.

23. Murphy, *Seizing the Means of Reproduction*, 87.

24. Murphy, *Seizing the Means of Reproduction*, 97.

25. Murphy, *Seizing the Means of Reproduction*, 39.

26. J. Nelson, "Abortions under Community Control," 158.

27. A. Nelson, *Body and Soul*.

28. D. Davis, *Reproductive Injustice*.

29. Stephanie Schiavenato, "Birthing under Investigation," *Fieldsights*, May 1, 2020, https://culanth.org/fieldsights/birthing-under-investigation.

30. Hester, *Xenofeminism*, 73–74.

31. Nina Martin and Renee Montagne, "Nothing Protects Black Women from Dying in Pregnancy and Childbirth," *ProPublica*, December 7, 2017, www.propublica.org/article/nothing-protects-black-women-from-dying-in-pregnancy-and-childbirth.

32. Cottom, *Thick*; Linda Villarosa, "Why America's Black Mothers and Babies Are in a Life-or-Death Crisis," *New York Times Magazine,* April 11, 2018, www.nytimes.com/2018/04/11/magazine/black-mothers-babies-death-maternal-mortality.html.

33. Joan Mukogosi, "Establishing Vigilant Care: Data Infrastructures and the Black Birthing Experience," Data and Society report, July 10, 2024. https://datasociety.net/wp-content/uploads/2024/07/establishing_vigilant_care_report.pdf.

34. Carolyn Johnson, "Long Overlooked by Science, Pregnancy Is Finally Getting the Attention It Deserves," *Washington Post*, March 6, 2019, https://www.washingtonpost.com/national/health-science/long-overlooked-by-science-pregnancy-is-finally-getting-attention-it-deserves/2019/03/06/a29ae9bc-3556-11e9-af5b-b51b7ff322e9_story.html.

35. Andalibi, "Symbolic Annihilation through Design."

9

"So Creepy It Must Be True!"

Techno-Orientalism, Technonationalism, and the Social Credit Imaginary

Nosediving into Social Credit

In 2014 and with minimal international coverage, the Chinese State Council published the "Planning Outline for the Establishment of a Social Credit System," a dense, unspectacular document that renewed old and set forth new, broad goals for implementing a national credit system.[1] Two years later and mere weeks before the 2016 US presidential election, the acclaimed British dystopian TV anthology *Black Mirror* released its episode "Nosedive." The first episode of its third season, and the first produced by Netflix, depicts a society where individuals rate and are rated at every encounter, with their score determining access to privileges and vulnerability to punishments. Almost immediately, these two disparate texts—of statecraft and entertainment—became su-

Newsweek

U.S. World Business Tech & Science Culture Newsgeek Sports Health The Debate

WORLD

'Black Mirror' in China? 1.4 Billion Citizens to Be Monitored Through Social Credit System

BY CHRISTINA ZHAO ON 5/1/18 AT 10:57 AM EDT

WORLD CHINA BEIJING PRESIDENT XI JINPING BLACK MIRROR

FIGURE 9.1. Screenshot of an article from *Newsweek*, captured March 14, 2021

tured in the discourse of popular English-language media outlets.[2] Report after report on China's Social Credit System (CSCS) came to deploy images and facets of "Nosedive" to describe features like mandatory "citizen scoring" using advanced artificial intelligence (AI) and big data, features that simply did not and still do not exist as national policy (see figs. 9.1, 9.2, and 9.3). At the same time, these reports were cited in countless listicles as further confirmation of the show's uncanny prophetic quality, leading many to speculate on how social credit is coming for, or already hidden within, "our" supposedly more open, democratic technology.[3] By 2018, *Black Mirror* became the term most associated with CSCS when searching for it on Google, and remains there.

Sinophobic fears and inaccurate depictions of China are nothing new, nor is science fiction sold as forecast. Since the early twentieth century, when British author Sax Rohmer introduced the supervillain Dr. Fu Manchu in a series of novels that condensed white fears over the perceived threat of Chinese labor as culturally backward yet unemotionally cunning, through the 1980s rise of the cyberpunk genre, which frequently imagines a future after Japan overtakes the United States against the historical backdrop of a declining US automobile manufacturing sector, techno-Orientalism in science fiction commonly constructs caricatures of Asia and Asians that are simultaneously pre- and hypermodern in ways that reflect entangled anxieties over technology and labor. Hugo Gernsback, the man credited with popularizing the term *science fiction*, emphasized the genre's potential for "prophetic vision."[4] While the case of "Nosedive" and CSCS certainly perpetuate these legacies of techno-Orientalism and the marketing of prophetic fiction, they

Esquire SUBSCRIBE

A 'Black Mirror' Episode Is Coming to Life in China

People will be prevented from traveling on trains and planes based on their social credit scores.

By Gabrielle Bruney Mar 17, 2018

FIGURE 9.2. Screenshot of an article from *Esquire*, captured March 14, 2021

INSIDER Log in Subscribe

DOW +0.53% S&P 500 +0.65% NASDAQ 100 +1.05%

HOME > TECH

China might use data to create a score for each citizen based on how trustworthy they are

Clinton Nguyen Oct 26, 2016, 12:33 PM

David Dettmann/Netflix

POPULAR WITH SUBSCRIBERS

Inside Mailchimp's 'mass exodus' of women and people of color

FIGURE 9.3. Screenshot of an article from *Business Insider*, captured March 14, 2021

present some peculiar insights into the cultural politics of creepy sci-fi and the technonationalist historical present. Unlike in previous instances and eras, this techno-Orientalism arrives in the circulation of their popular readings and associations, rather than from their respective texts. "Nosedive" conspicuously lacks any Asian characters or aesthetics, let alone a single mention of "social credit," and CSCS as an actual policy is remarkably unremarkable, lacking any flashy AI or nationally mandated "citizen scores." So how did this episode come to serve as the narrative representation or proof of a particular imaginary of CSCS that threatens to creep into the United States, rather than as a critique of US technology platforms themselves and their dependence on cheap labor and financial credit?

Addressing this question demands a turn from what makes a technology creep(y) to how technologies, their systems, and where they are from come to be read as creepy and provoke anxiety and for whom. Indeed, part of the problem is just how often we incorrectly intuit inanimate technologies as creeping, as moving autonomously and crawling with their own agency. What is at stake here is not whether CSCS or *Black Mirror* are themselves creepy. Instead, we must reckon with how histories of race and labor (i.e., who makes and is made to make what on this planet, under what conditions, and how those relationships are made to appear natural) slip undetected into cultural readings of some technologies as more open or closed in the context of the fraught relationships between China, the United States, and their respective "big techs."

That "Nosedive" was released at the onset of a historical conjuncture commonly referred to as the "techlash," when the US tech industry came under increased public scrutiny following Brexit, the US election, and subsequent revelations of manipulation and abuse of user trust, is central to this story. *Black Mirror*'s reputation as a creepy show that aimed to both warn *and* prophesize was already well established, which immediately conferred credibility to paranoid fantasies and misleading comparisons of CSCS and "Nosedive." Meanwhile, personal networked devices and social media platforms once accompanied by wonder or ambivalence before late 2016 became increasingly recognized as intrusive and felt to be creepy. In turn, China's very absence in "Nosedive" contributes a key ingredient in the credible creepiness of what I call the "social credit imaginary." The indirectness not only lent legitimacy to this reference because the fan/news consumer was empowered to make it, but it also served to displace real concerns over big tech's investments in the financialization of everyday life (e.g., subtly scoring, nudging, and monitoring user behavior) onto China. Even as "Nosedive" has faded into the technocultural background, features of the "Nosedive" society continue to shape how CSCS is misapprehended, particularly the notion that the very concept of social credit is Chinese and that it is coming for, or already in, the United States. As I will demonstrate, this imaginary, and not research into the existent system or genuine interest in China as a place in contemporaneous temporality, racializes some creepy technologies and the very concept of control as Chinese and set in a future the United States and the West must avoid. Rather than the actual CSCS or even "Nosedive" itself, I contend that it is the *popular misreadings* of "Nosedive" and this social credit *imaginary* that should be recognized as truly creepy. That is, this specter of Chinese social credit provides threateningly creepy yet navigable and pleasurable aesthetics of the near future by feigning a revelation of the late capitalist logics that lie behind our ubiquitous black mirrors. As Jeremy Daum, one of the leading experts on Chinese law, puts it, "There are two stories worth exploring [with CSCS]; what is actually happening in China and what we fear is happening to ourselves."[5] Plenty of experts have worked tirelessly to dispel the misconceptions around the former; this chapter is primarily concerned with the cultural politics of the latter.

The Pleasure of Prophecy and Technocreep as a Contemporary Micro-Sci-fi Genre

Few contemporary cultural artifacts better epitomize technocreepy aesthetics and tropes than *Black Mirror*. Rhyming with Thomas Keenan's concept of technocreep as "a one way trip to the total surrender of privacy and the commodification of intimacy," which this anthology serves to complicate, the showrunner aimed to warn audiences of "the dark side" of technology and what would become of the future "if we're clumsy."[6] The show's title makes obvious reference to the off screen of a smartphone, with its blackness simultaneously blocking the symbolic recognition of a user's real relationship to their device and offering a moment of reprieve from the overstimulation of the apps it runs. Over the last decade, the show's reception and interpretations have mostly tracked the prevailing attitudes toward US big tech. Its first two seasons and Christmas special ran on the UK's Channel 4 from late 2011 through 2014 at a time when techno-utopianism still prevailed. By early 2016, *Black Mirror* was a touchstone in the US pop technocultural vocabulary, with *Black Mirror-esque* and its variants becoming placeholders for an unnamed, intimate, and invasive quality of new gizmos' and gadgets' promiscuous data practices and behavior modification techniques.[7]

Black Mirror ✓ N
@blackmirror

This isn't an episode. This isn't marketing. This is reality.

7:28 PM · Nov 8, 2016

105.2K Reposts **5,482** Quotes **128.6K** Likes **46** Bookmarks

FIGURE 9.4. On US election night 2016, *Black Mirror*'s Twitter account responded to a flurry of tweets from fans like "I don't like this episode of black mirror" and "what is happening why is every broadcast tv channel showing the same episode of black mirror."

At the same time that the show embraced its status as a warning, it also invited its fans to draw parallels to the real world and achieved a level of credibility from appearing to prophesize the future as a result. In a historical moment of increasing precarity, there is a strange comfort and consumer pleasure produced when seeing one's fears about the future confirmed, and fiction that exhibits privileged access to what lies beyond the horizon of the present has certain cathartic appeal. Such an appeal quenches that thirst for a satisfying "I told you so!" or "I knew there was a creep behind me all along!" Propelling the show's cultural circulation through a *sensation* of prophecy, listicles abound with headlines like "The 7 'Black Mirror' Prophecies That Have Come True" along with hyperreal marketing that feeds on this reputation (see figs. 9.4 and 9.5).[8] *Black Mirror*'s creator, Charlie Brooker, further fueled this flame when he publicly predicted both Brexit and Trump's electoral victory.

It is against this backdrop that "Nosedive" must be situated. Set in a retro-futuristic, pastel, suburban landscape, the episode follows Lacie (Bryce Dallas Howard), a white, single, middle-class working woman, as she navigates a society that looks as if it was designed as a collaboration between Facebook, Uber, and Yelp.[9] Everyone is assessed by everyone else, resulting in a cumulative score on a five-point scale that affects and is affected by one's social network. The higher one's score, the more weight one has when rating others. Characters use a smartphone-like device connected to an augmented-reality-enabled contact lens that shows others' scores above their heads. The narrative hinges on Lacie's attempt to increase her score in order to secure a discounted rental rate at a fancy apartment; she tries to do this by garnering enough five-star ratings from "high fours" at her childhood friend's wedding. As the title suggests, this does

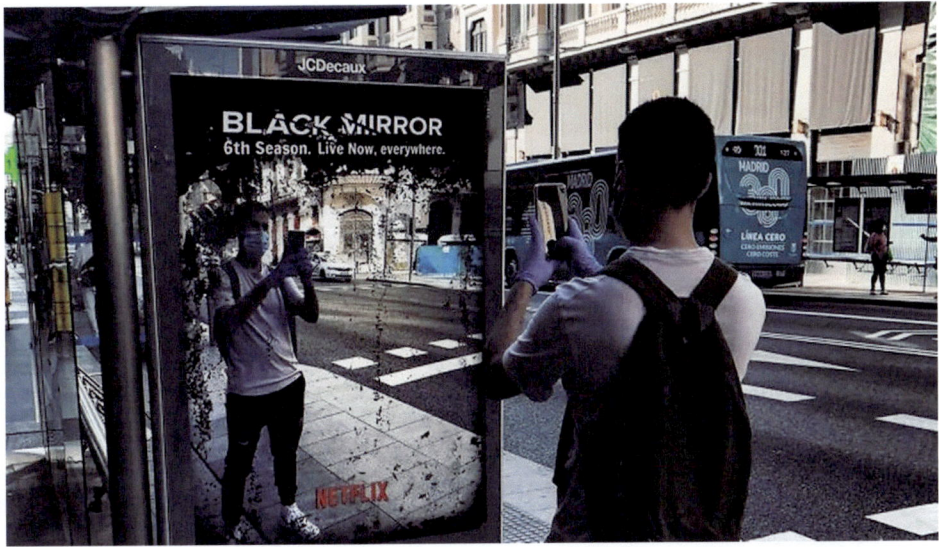

FIGURE 9.5. An ad in Madrid promoting a nonexistent sixth season of *Black Mirror* during the summer of 2020. There was no sixth season on the horizon at the time. Brooker remarked that he doubted the public could "stomach" another season during a pandemic.

not go as planned. The episode unfolds as a series of unfortunate events, each resulting in snowballing point deductions and further obstacles along Lacie's trip to the wedding reception. The harder she tries to raise her score, the harder and more precipitously it falls.

Following its release, the commentary mostly sidestepped the episode's obvious connections to the cultural effects and disproportionate harms of rating systems across lines of race, class, and gender developed by US tech platforms to "disrupt" middlemen and traditional regulatory institutions. In a promotional interview for the third season, Brooker is asked about CSCS: "I did see that. . . . They're going to do the system from 'Nosedive' for citizens! It's incredibly sinister."[10] In a separate interview, episode cowriter Rashida Jones recounts: "Charlie Brooker is like some kind of weird prophet, everything he's ever done has come true. . . . Even while we were working on 'Nosedive' he sent us an article about what was going on in China, where there was like a full personality rating system that could impact your ability to travel on public transportation, to get a job, and get housing, really, really intense things that should not be real—and now they're real."[11]

The day after the episode's release, *Black Mirror*'s own Twitter account retweeted a *Washington Post* article about CSCS in the form of a tongue-in-cheek

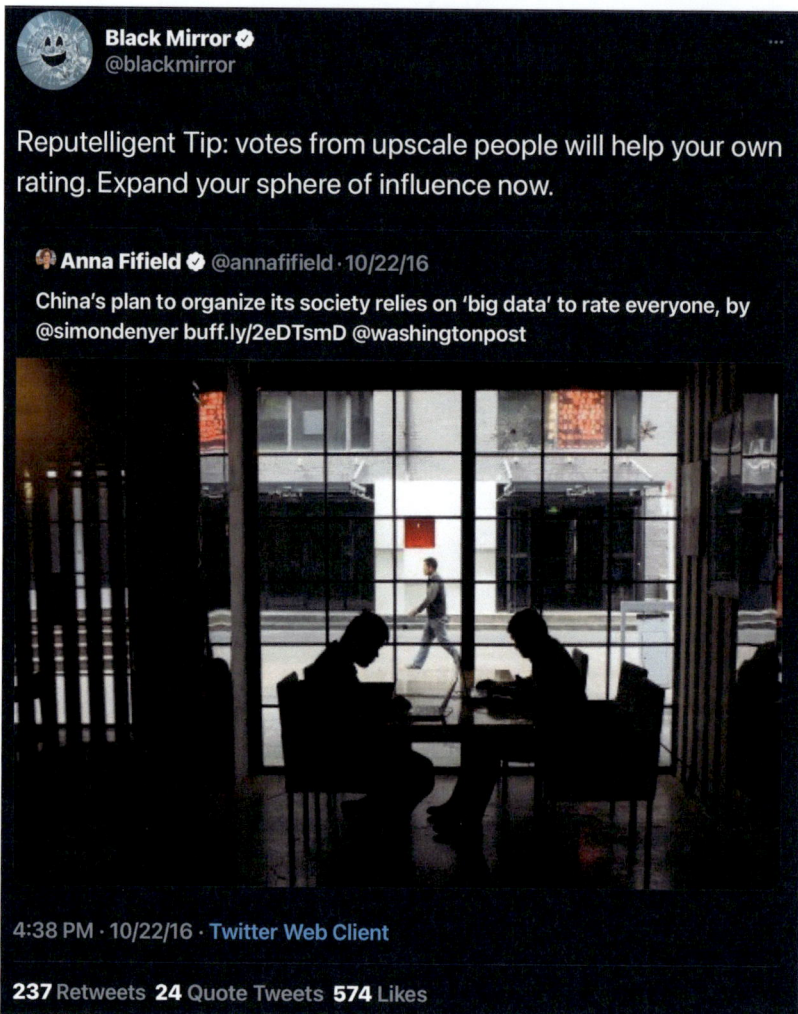

FIGURE 9.6. *Black Mirror*'s own Twitter account uses a fictional company from "Nosedive" to misleadingly retweet a report from the *Washington Post* about CSCS.

advertisement for the fictional reputational consulting firm from "Nosedive," Reputelligent (see figs. 9.6 and 9.7). Maintaining its own prophetic status is hardly incidental. It is a core component of its persistence in popular techno-cultural vocabulary, lending *Black Mirror* its own form of social credibility in an age of uncertainty. Once it was a surprise for life to imitate art; today it must. To maintain its credit/credibility, real-world proof must be issued to pay off the debt of its predictions.

FIGURE 9.7. In "Nosedive," Lacie meets with a Reputelligent consultant to learn how she can achieve a 4.5 score.

The Social Credibility of Projection

The association of "Nosedive" with CSCS was likely reinforced through the visual resemblances of app interfaces. The US press caught wind of a Chinese app called Sesame Credit, run by Ant Financial, a subsidiary of the e-commerce giant Alibaba. Sesame Credit was launched based on goals from the Chinese State Council's planning outline for CSCS. Rather than use a five-point scale as in "Nosedive," Sesame Credit's home screen contained a three-digit number up to eight hundred, similar to a US credit score, with several gamified designs to encourage engagement (fig. 9.8). At that time, Beijing issued permits to several municipalities and eight tech firms, including Alibaba, to pilot social credit systems.

The Sesame Credit app algorithmically scored users on several inputs but mostly relied on purchasing habits made using Alibaba's electronic payment platform, Alipay—a detail often ignored in coverage. Obviously, there are some similarities to Lacie's world. In both, apps were deployed with scores that nudge user behavior through ratings and incentives. But that is effectively where the similarities end. For one, Chinese user ratings do not affect others' scores. Sesame Credit functions far more like an opt-in, glorified loyalty rewards program than a mandatory credit score, one that is limited to Alibaba's and its partners' data ecosystems and is *not* part of any national credit system. Both Lacie's and Sesame Credit's apps seem to offer to casual US consumers of

FIGURE 9.8. The Sesame Credit app interface (photo via AP/Imagechina)

news about technology and China what these consumers so often lack in real life: a visualization of the rules of the game, however distorted, of what determines their own credit and therefore social prospects. This creepy conflation of fiction and fact displaces the massive, unregulated, shadow market of US-based data brokers and platforms that produce a variety of "e-scores" as well as mandatory credit scores for US residents, by reassuring audiences that *at least it is worse over there*.[12]

The Chinese government eventually pulled the plug on the experiment with the tech firm because of the conflicts of interest in letting companies like Alibaba, which sell financial services and consumer goods, also determine credit scores. Despite this crucial update, most coverage of CSCS continues to rely on the app interface, its three-digit score, news about blacklists, and stock images of facial recognition feeds from CCTV cameras to imply that a version of Sesame Credit would eventually become mandatory nationwide and integrated into every aspect of the state's surveillance apparatus.[13]

Still, these few aesthetic similarities between Sesame Credit's and Lacie's apps appear to have been enough to tap into and fuel larger fantasies of a dystopian Chinese digital dictatorship using some of the very algorithmic dataveillance mechanisms integral to Silicon Valley platforms. This social credit imaginary serves as a convenient foil of control to the presumed freedom of Western tech. We could summarize the predominant logic thus: because China is blanketed in surveillance technologies, with social media censored, all matter of communications heavily monitored, CCTV cameras everywhere hooked up to real-time facial recognition software, and every purchase tracked, it must therefore be hoovering up all that data into a central system. And, since the government is authoritarian, it must be feeding it into a massive black box so that when you do something that indicates to this mysterious system that you or your friend violated a law or a moral principle or are a bad communist, it triggers an automated process to ding your single, individual, mandatory social credit score and potentially block you from social privileges. It feels like a "digital dictatorship" where big data meets Big Brother. It *feels creepy*.

There are indeed elements of truth to this. The country is one of the most, if not *the* most, intensely surveilled in the world, with cases of repression worth raising an alarm over—Xinjiang and Hong Kong, to name only two. And surely the Chinese state can obtain and censor nearly any piece of information within its borders. However, reading translations of the five-year plan or subsequent CSCS legal documents discloses a conglomeration of disparate goals and initiatives that are far less thrilling and monolithic than what English-language media imagine and perhaps fantasize. Daum frequently shares how at nearly every speaking engagement he must spend far more time dispelling myths about CSCS than describing what it actually is.[14] That Beijing has yet to comment on these inaccuracies, let alone the bad press, given how frequently it seeks to rebuke or spin reports, should at minimum give anyone who is warning the world about the looming dangers of CSCS pause. *Social credit* has incorrectly become a synonym for all Chinese high-tech surveillance technologies and censorship regimes. In reality, CSCS mostly consists of strategies for increasing the public's trust in institutions within a market economy that *as a consequence* has produced rampant fraud, determining financial credit for individuals who have no lending history, and installing information-sharing mechanisms for purposes of enforcing existing laws. Without going into the intricacies of CSCS, the following are five of the most common misconceptions spun in English-language popular coverage outside China over the last few years.

First, while a few regions and tech firms experimented with different individual social credit rating schemas, there is *no* single, all-encompassing score or rating system for everyone in China.[15] Second, *social credit* (社会信用, *shehui xinyong*) holds very different connotations in Chinese; the term evokes the notions of "public trust" and "honesty" rather than "credit" determined by one's "social media."[16] Third, CSCS is far more concerned with the social credibility of institutions and firms than individuals.[17] Fourth, there is no unified, standard system. CSCS is more accurately a flexible "umbrella category encompassing several moving parts of a broader policy agenda that includes both national initiatives as well as city-level pilot projects that do not generalize to a countrywide scale."[18] And finally, fifth, its operations are remarkably low-tech: more data sharing, administration, publicity, and basic math than AI or advanced algorithmic governance. Reports consistently conflate CSCS with other examples of surveillance technologies in China, like facial recognition. There is certainly a vast surveillance network with many high-tech tools, though there is no direct integration between the CCTV network and the social credit system. It's far more just a set of super–Excel documents than a superalgorithm.[19]

There are plenty of reasons to be suspicious of the Chinese Communist Party's (CCP's) claims and watchful for abuse and potential harm. To what extent these initiatives work as intended and how they are assessed is undoubtedly up for further investigation and debate. Yet too often, reports take the premise of "Nosedive" and its cursory visual similarities to Sesame Credit as sufficient confirmation of the reality of CSCS. They take CSCS as a mandatory, nationwide version of a Silicon Valley platform. The more important question lies in how these misunderstandings register ideological solutions to real-world contradictions of digital technology under late capitalism.[20] In other words, these are more than mere inaccuracies—they make up the contours of a politically compelling, spectral foil.

The Creep of Technonationalism

Technology does not creep, at least not on its own. It is easy to think that it does. "Technology" as a concept has long been misunderstood as autonomous, itself an agent of change—*changing the way we live*.[21] The etymology of *creep* summons the Old English verb *creopan*, noting the strange movements of reptiles and insects. Its noun, *creepiness*, elicits the unrecognizable, frightening, and felt qualities of that crawling, particularly on or beneath one's skin. Today, technocreepiness evokes a sensed, not seen, lurking presence within or behind these devices with nefarious goals threatening to invade bodily or psychic autonomy. There are good materialist accounts for this impression. The utopian visions of advances in information and communications technology reversing the economic stagnation faced in the capitalist core after industrial production moved largely to China have revealed themselves as empty and misguided.[22] Users in the Global North are increasingly drawing connections between ubiquitous consumer surveillance schemes seeking to make intimate predictions about them and US tech platforms' massive profits. This technocreepiness is an affective symptom of these present investments in technologies of control after the veneer of Silicon Valley's newspeak of "disruption," "sharing," "participation," "democratization," and "revolutionizing the way you [fill in the blank]" has cracked.

Those early visions entailed promises for transcending borders and geopolitics, particularly for the emergence of a new global superpower. Many in the early decades of the consumer internet were convinced the internet would democratize China. Bill Clinton infamously equated Beijing's early efforts to censor the internet to "nailing Jell-O to the wall." Now, the script has been flipped. That inevitable dream has been superseded by panic that China is changing "our" internet. With big tech facing countless systemic and PR crises, China has slipped into a new role as a convenient ideological foil both to explain why Euro-American technoliberal visions of a free and open, borderless medium of information and financial exchange have failed and to falsely justify US tech firms' market dominance for fear that Chinese technological "values" will creep into "ours."

After years of learning Mandarin, inviting Chinese officials to his company headquarters, and even going on an infamous "smog jog" through Beijing's Tiananmen Square to convince the Chinese Communist Party to let Facebook do business in the country, Mark Zuckerberg has adopted a "technationalist" tune.[23] In a 2019 speech, Zuckerberg professed: "China is building its own internet focused on very different values, and is now exporting their vision of the internet to other countries. Until recently, the internet in almost every country outside China has been defined by American platforms with strong free expression values. There's no guarantee these values will win out. A decade ago, almost all of the major internet platforms were American. Today, six of the top ten are Chinese. . . . Is that the internet we want?" Put differently, *you may not like us anymore and we may be a little evil now, but at least we're not China.*[24]

Even Tristan Harris, the so-called conscience of Silicon Valley who has built a career ostensibly opposing Facebook, has joined in on the Sinophobic drumbeat.[25] In April 2021, Harris, cofounder of the Center for Humane Technology and star of the 2020 Netflix documentary *The Social Dilemma*, was invited to speak before the US Congress for a second time in three years. There Harris argued that social media poses a threat to the US state on par with Cold War nuclear standoffs for the way its business model has "degraded the capacity of the American brain." For him, this stands in the way of addressing "genuine existential threats," of which he lists "the rise of China" ahead of the climate crisis and inequality. Harris concludes his prepared remarks by calling for a "humane, clean 'western digital infrastructure'" before warning that "today we are offered two dystopian choices: either to *install a Chinese 'Orwellian' brain implant into society with authoritarian controls, censorship and mass behavior modification. Or we can install the U.S./Western 'Huxleyan' societal brain implant* that saturates us in distraction, outrage, trivia and amusing ourselves to death. Let's . . . encourage a 3rd way, to have the government's help in incentivizing Digital Open Societies worth wanting, that *outcompete Digital Closed Societies*" (emphasis added).[26]

George Orwell versus Aldous Huxley. China versus the United States and "the West." A clash of civilizations at the site of the brain—not an individual but a national brain. This renewed technonationalism arrives cloaked in a false neutrality. The "US/Western 'Huxleyan' societal brain implant" is framed as an aberration from its true "openness" with a potential for renewal, whereas the "Chinese 'Orwellian'" one is always already authoritarian. Sounds like dystopian science fiction. Sounds creepy, sounds true.

These alarms over looming East Asian technological threats stand as just the latest iteration of a legacy of white nationalistic fears of replacement and techno-Orientalist tropes in the United States. Still, this chapter ventures to demonstrate how the remarks of both the CEO and his critic are emblematic of a new layer of Sinophobic techno-Orientalism in the historical present that ideologically papers over contradictions confronted by US tech platforms as the public relations and economic success to which they have grown accustomed are threatened with structural crises. In contrast to previous eras, this discourse has turned to racializing ambiguous notions of the brain, society, and technology without necessarily referencing Chinese people or places. Instead, now particular technologies and their presumed political inclinations are themselves coded as controlling and thus Chinese, understood as an implant, lying beneath the skin, invisible to the naked eye, coming for and already residing in American brains and society, operating according to incomprehensible logics of control. We are thus left with the tautological conclusion that this rhetorical societal brain "implant" is Chinese because it is Orwellian and is Orwellian because it is Chinese. For all the panic over polarization today, Sinophobia and general skepticism toward technology platforms represent two of the few widely shared sentiments across US society. *China is no longer just taking "our" jobs; China is taking "our" phones and "our" brains. And only Silicon Valley and its "open" values can save "us."*

NOTES

1. The platform China Law Translate has produced a translation of the initial planning document; see "Establishment of the Credit System," China Law Translate, April 27, 2015, www.chinalawtranslate.com/socialcreditsystem. For a summary of the document, see Jeremy Daum, "'Map' of the 2014–2020 Social Credit Plan," China Law Translate, De-

cember 25, 2017, www.chinalawtranslate.com/map-of-the-2014-2016-social-credit-plan. For the original document, see "国务院关于印发社会信用体系建设规划纲要（2014–2020年）的通知_政府信息公开专栏," Chinese State Council, accessed April 22, 2022, www.gov.cn/zhengce/content/2014–06/27/content_8913.htm.

2. These reports were predominantly found in US outlets, but Canadian, British, and Australian outlets also published similar pieces.

3. This fear manifested among several public figures. Both liberal billionaire George Soros and former vice president Mike Pence explicitly identified CSCS as a threat to the United States. Right-wing media personalities found the techno-Orientalist and racist tropes of "becoming like China" particularly useful in opposing COVID-19 public health measures, social media moderation, and corporate responsibility initiatives. For instance, in a November 2021 podcast episode, the dubious and controversial Joe Rogan remarked: "That's what terrifies me. We have to become like China in order to deal with what they're doing. I just feel like one step moving in that general direction is a social credit score system and I'm terrified of that, and I think that is where vaccine passports lead to, I really do." Joe Rogan, "#1736–Tristan Harris and Daniel Schmachtenberger," *Joe Rogan Experience*, podcast, November 18, 2021, https://youtu.be/im4O2sW3FiY.

4. Sherryl Vint reminds us that Gernsback made the remark in marketing his magazine, *Amazing Stories*. Vint, *Science Fiction*.

5. Jeremy Daum, "China through a Glass, Darkly," China Law Translate, December 24, 2017, www.chinalawtranslate.com/china-social-credit-score.

6. Keenan, *Technocreep*; Charlie Brooker, "The Dark Side of Our Gadget Addiction," *Guardian*, December 1, 2011, www.theguardian.com/technology/2011/dec/01/charlie-brooker-dark-side-gadget-addiction-black-mirror.

7. One could easily argue such terms have become the rough equivalent of *Orwellian* for the twenty-first century.

8. Brock Wilbur, "The 7 'Black Mirror' Prophecies That Have Come True," *Inverse*, May 13, 2016, www.inverse.com/article/15540-how-many-episodes-of-black-mirror-have-already-come-true.

9. An underappreciated aspect of the show's transition from a UK to a US production was a notable shift in theme. Apart from "White Bear," the second episode of the second season, the episodes prior to "Nosedive" relied heavily on themes of compromised or humiliated masculinity in relation to advanced technologies, as well as an implied nostalgia for "simpler times." For instance, in "National Anthem" the UK prime minister, under pressure from social media polls, is forced to have sex with a pig on live TV by a terrorist/performance artist who took a princess hostage. In the climax of "15 Million Merits," the male protagonist's grand ethical stand opposing the alienating conditions of his life is co-opted and repackaged as a media spectacle. "The Entire History of You" imagines a world with brain implants that store every memory, and the narrative ends up confirming the lead's misogynistic paranoia of his wife's infidelity.

10. Marlow Stern, "'Black Mirror' Creator Charlie Brooker on China's 'Social Credit' System and the Rise of Trump," *Daily Beast*, October 27, 2016, www.thedailybeast.com/articles/2016/10/27/black-mirror-creator-charlie-brooker-on-china-s-social-credit-system-and-the-rise-of-trump.

11. Jude Dry, "Rashida Jones Links 'Black Mirror' Episode She Wrote to Her Social Media Anxiety," *IndieWire*, May 2, 2019, www.indiewire.com/2019/05/rashida-jones -black-mirror-social-media-1202130483.

12. O'Neil, *Weapons of Math Destruction*, 149.

13. Blacklists of various kinds in China have attracted media attention as well; among other punishments, they prevent the purchasing of high-speed train tickets. While they indeed fall under the social credit score umbrella, the blacklists do not consist of people with low scores nor are they related to the Sesame Credit program. Instead, those placed on national or regional blacklists end up there because of intergovernmental information sharing and the enforcement of court orders.

14. Daum is a senior research fellow at Yale Law School's Paul Tsai China Center and runs the China Law Translate platform, which translates and hosts an archive of every government document associated with CSCS. See "Social Credit Resources Page," China Law Translate, accessed January 5, 2024, www.chinalawtranslate.com/en/social-credit -resources.

15. 2020 came and went without the compulsory social credit scores nearly every news report cited as inevitable, mistaking a pilot program or a tech firm's marketing for national policy. That policies could look different across different regions and that a firm's interests may not be in lockstep with the state is incomprehensible to many outside reporters. Instead, CSCS calls for registering universal social credit *codes*, mostly to public and private organizations for purposes of tax collection and law enforcement across siloed government agencies. Companies may be ranked or scored by an industry association, e.g., on food safety, which is of widespread concern in China; however, CSCS itself does not produce them. In specific and relatively serious offenses, individuals and companies may be blacklisted after they are convicted of a crime and fail to follow a court order, but again, this is not the result of a bad score. See Louise Matsakis, "How the West Got China's Social Credit System Wrong," *Wired*, July 29, 2019, www.wired.com/story/china-social -credit-score-system; Jamie Horsley, "China's Orwellian Social Credit Score Isn't Real," *Foreign Policy*, November 16, 2018, https://foreignpolicy.com/2018/11/16/chinas-orwellian -social-credit-score-isnt-real; and Shazeda Ahmed, "The Messy Truth about Social Credit," *Logic(s)*, May 1, 2019, https://logicmag.io/china/the-messy-truth-about-social-credit.

16. Especially in the context of reporting on digital technology, English-language readers might associate social credit with social media or interpersonal relations. Particularly pertinent to this chapter, this connotation was further reinforced in reporting that cited Sesame Credit's marketing claim that one's friends' scores would affect one's own, later finding resonance with Lacie's *Black Mirror* world.

17. In fact, unless an individual owns a business, is a public official, has taken out a loan or credit card, has been convicted of a crime, or has defaulted on a court order, most citizens are not included in this nationwide database. Since China began its economic liberalization under Deng Xiaoping in the late 1970s, the country has struggled with pervasive corruption and fraud arising out of market liberalization, "free" of consumer safety protections and reliable financial regulations. CSCS is much more concerned with enforcing existing consumer safety laws and regulating institutions both public and private. This severely underappreciated aspect should be of great interest to the United States,

especially considering its widespread domestic crisis of trust in institutions as a result of its own neoliberal policies. Horsley, "China's Orwellian Social Credit Score Isn't Real."

18. CSCS might be better understood as an amalgamation of systems rather than a singular, coherent apparatus. That China could be politically and technologically decentralized is incongruous with many depictions of the country. Moreover, reporters would do well to distinguish between the marketing copy of Chinese tech firms, who (as in the United States) overpromise in hopes of wooing private and/or state investments. Shazeda Ahmed, "Messy Truth about Social Credit."

19. E.g., if one has a court order against them, they may be blacklisted from buying high-speed-train tickets because of CSCS's information-sharing system.

20. Vint, *Science Fiction*.

21. Winner, *Autonomous Technology*, 9; Marx, "'Technology.'"

22. Smith, *Smart Machines and Service Work*.

23. On the "smog jog," see Walker Benjamen, "Zuckerberg in China," *Zuckerberg Review*, 2018, http://zuckerbergreview.com/vol_2/walker.html. While Facebook's apps are not accessible to users living in mainland China without a virtual private network, Facebook does still operate in China, operating platforms and offices for Chinese firms to buy advertisements on their apps. Shoshana Wodinsky, "Facebook Says China Is Its Biggest Enemy, but It's Also a Highly Valued Customer," *Gizmodo*, July 29, 2020, https://gizmodo.com/facebook-says-china-is-its-biggest-enemy-but-it-s-also-1844526005.

24. In a similar vein, Google offered to build a censored version of its search engine for the Chinese market before facing backlash from its employees and US legislators. In a summer 2020 congressional hearing, Google's CEO, Sundar Pichai, defended the company against accusations of working with the Chinese by citing a recent enormous contract it signed with the US Department of Defense. J. S. Tan, "Big Tech Embraces New Cold War Nationalism," *Foreign Policy*, August 27, 2020, https://foreignpolicy.com/2020/08/27/china-tech-facebook-google.

25. Bianca Bosker, "What Will Break People's Addictions to Their Phones?," *Atlantic*, October 8, 2016, www.theatlantic.com/magazine/archive/2016/11/the-binge-breaker/501122.

26. Quoted from submitted written testimony, which differs from Harris's spoken testimony, as is sometimes the case. Tristan Harris, "Algorithms and Amplification: How Social Media Platforms' Design Choices Shape Our Discourse and Our Minds," statement submitted to the US Senate Committee on the Judiciary, April 27, 2021, www.judiciary.senate.gov/imo/media/doc/Harris%20Testimony.pdf.

It should be noted that this dichotomy and language is borrowed directly from Neil Postman's introduction to *Amusing Ourselves to Death*, a book Harris frequently cites and reads from in interviews. See Ezra Klein, "How Technology Brings Out the Worst in Us, with Tristan Harris," *Ezra Klein Show* (Vox Conversations), podcast, February 19, 2018, https://www.vox.com/technology/2018/2/19/17020310/tristan-harris-facebook-twitter-humane-tech-time, accessed July 28, 2924; and Joe Rogan, "#1558–Tristan Harris," *Joe Rogan Experience*, podcast, October 30, 2020, www.youtube.com/watch?v=OaTKaHKCAFg.

10

Resistant Resonances

Vocal Biomarkers, Transductive Labor, and the Politics of Things Not Heard

"So, do you think the computer said he's depressed?" asks Klaus.

I am sitting with Klaus, a White, European engineer in his thirties, in his office at a West Coast university in the United States.[1] He has just shown me a video of an avatar guiding a mental health-care assessment interview of a human research subject, selected from the dataset of a study that he and his team completed in 2015. The avatar is the user interface of the study's final product: an automated system designed to screen a person for depression and posttraumatic stress disorder (PTSD) based on how they sound rather than what they say. This system has drawn me, in the winter of 2017, to join Klaus's team as a research assistant–ethnographer in pursuit of a larger project on the impact of algorithmic listening technologies on the US mental health-care system. Like the two other teams with which I conducted ethnographic fieldwork,

Klaus's is part of a growing effort to utilize signal processing, machine learning, and other data-driven techniques of pattern recognition to seek out "vocal biomarkers," or biological markers of mental distress supposedly resident in the sonic contours of the voice.

Taking up Klaus's query despite my lack of clinical training, I try to disentangle the semantic meaning of the subject's speech from the pacing, intonation, and breathiness of his voice, the sorts of features I imagine the system would analyze. I make a guess: he sounds depressed. Incorrect, according to the "computer," says Klaus. As Klaus cues up another video, continuing his demo of the dataset (the interviews) that forms the backbone of the system's software, I begin to feel as if I am also witnessing a demo of the limitations of my listening compared with the system's artificial intelligence (AI)–enabled software.

At face value, my initial impression of Klaus's technology seems to reiterate the central thesis of the corporate, academic, and federal actors invested in vocal biomarkers, who cite the sensory superiority of AI as its most alluring feature. In theory, vocal biomarkers are anchored in the biological mechanisms of mental distress and are expressed involuntarily in the sounds of a person's voice whenever they speak. However, they are also supposedly "human-imperceptible" due to their subtlety and lack of connection to semantic meaning, sociocultural convention, intentionality, and the social and political context of the clinical worker–patient encounter.[2] Supporters of vocal biomarker initiatives likewise assert that AI is fractured from and diametrically opposed to the human, rendering AI uniquely capable of pinning down vocal biomarkers in streams of speech. Taking for granted the matrices of empire, race, gender, and capital that continually structure and refract through the human-machine distinction, vocal biomarker stakeholders therefore argue that AI, unlike the overly subjective mental health-care worker, can perform the narrow, biologically essentializing listening that audializing vocal biomarkers requires.

Many mental health-care professionals thus see vocal biomarkers (and the AI deemed necessary to pin them down) as the antidote to Western psychiatry's gravest weakness: its reliance on communicative interactions, rather than biological assays, to gauge a person's psychological state. As one neuropsychiatric researcher pursuing vocal biomarker research explains, "In psychiatry, we don't even have a stethoscope. . . . It's 45 minutes of talking with a patient and then making a diagnosis on the basis of that conversation."[3] To its supporters, vocal biomarker research entails the creation of a stethoscope for conversation itself, a technique for transforming the voice into a "vital sign" so that clinicians might listen directly into the beating, biological pulse of psychosocial disabilities.[4] Re-mediating the voice into a "window into your body" will purportedly enable clinical workers to discern the mentally ill from the mentally well, efficiently sorting those who are in need of mental health care from the less deserving, and the docile from the dangerous.[5]

Meanwhile, a growing chorus of critics warns that vocal biomarker technologies bespeak an ever-expanding "panaudicon": a totalizing surveillant ear that blurs the boundary between public and psychic life, marking the encroachment of AI into the intimate and otherwise inaccessible sphere of the mind.[6] These critiques, while important, tend to align with the dominant framework of technocreep that this volume is dedicated to recalibrating. They primarily characterize the harms that vocal biomarker technologies portend through binaries of privacy and surveillance while depicting users and makers of vocal biomarker technologies as uncritical and agentless. For instance, a 2021 editorial in *Lancet Digital Health* questioning the clinical utility of vocal biomarkers cautions that "voice data are personal data that could be used to reveal a person's gender, age, language, geographical and socio-cultural origins, and health."[7] This critique imagines a normative user/patient who has never before experienced being interpolated by a White, colonial "hegemonic perceiving subject" that rigidly maps them onto social categories based on the sound of their voice, someone who has never been told that they "sound," for instance, disabled, trans, or not American.[8] Other critiques portray vocal biomarker technologies as creepy because they can plumb the depths of a speaker's psyche without the speaker ever knowing, making it "nearly impossible for people to hide their feelings" and turning even the unintentional expression of affect into a potential mechanism of control and commodification.[9] While important, many of these critiques subtly direct scrutiny away from the very same technoscientific claims of vocal biomarker technologies that the more techno-optimistic characterizations turn on, or the notion that the technologies can reveal objective facts about a person's psychological or affective state. In so doing, these critiques leave less room for exploring how the intended targets of vocal biomarker technologies, and the people involved in developing them, might subvert or strategically reconfigure the facts that vocal biomarker technologies are designed to produce.

How might we "reclaim the messy contradictions," as the introduction to this book suggests, of the technologically mediated encounters that wrap around these technologies, moving beyond dominant accounts of vocal biomarker technologies as either promising public health intervention or perversely expanding algorithmic surveillance? Following this volume's provocations to pursue technocreep as a method, could the "creepiness" of vocal biomarker technologies be reimagined as generative of surreptitious, counterhegemonic values and relations? This chapter addresses these inquiries through ethnographic fieldwork conducted alongside researcher assistants and human research subjects, whose interactions are crucial to assembling clinical algorithmic listening initiatives. I argue that many of the mainstream analyses of this emergent field of research and technology are predicated on the notion that vocal biomarkers can and do faithfully represent interior, psychological states. Investments in the technology's ontological stability foreclose on more nuanced readings that surpass frameworks of efficiency, biological normativity, and privacy.

To analytically expand the horizon of what vocal biomarker technologies might make possible, I tune in to the individuals often unmentioned in discussions of vocal biomarker technologies' promises or perils and that I myself had not listened carefully for in Klaus's office: the interns and research assistants creeping at the edges of the video frame managing research subjects' interactions with the avatar, and the research subjects themselves. Focusing on their contributions underscores that vocal biomarker technologies are enacted through a set of contingent, sonic-communicative practices I call transductive labor, or a form of maintenance work that involves moving and transforming sound across multiple media. Research personnel, ranging from salaried employees to undergraduate interns, engage in transductive labor by eliciting, recording, and labeling vocalizations from research subjects. The research subjects engage in transductive labor by producing speech in exchange for a cash stipend. Their role is to provide the starting point of an algorithmic system that will calculate the resemblance between their own voices, embedded in the system's dataset, and the voices of future, anticipated speakers. Along the way, as research personnel and the subjects interact, listen, and negotiate the initial, sonic inputs at the front end of the team's transductive chain, they hit snags, or they collaboratively craft bursts of catharsis and care that do not fit neatly into normative biological imperatives of intervention and cure. These things not heard—the import of transductive labor to the overall production of vocal biomarker research, and the absences, additions, and nonhegemonic care practices this form of labor generates—complicate claims of the technology's expansive auditory abilities. In this way, homing in on transductive labor upends the hoped-for and feared capacity of AI to fully capture mental distress through the voice.

While the field of machine vision boasts a variety of publicly available image datasets that can be used to train an algorithm, no corollary exists for sound-based datasets, let alone sound linked to psychiatric unwellness.[10] Vocal biomarker researchers must often gather and label their own datasets, resourced from human subjects who live with or have been diagnosed with mental distress.[11] At its core, then, vocal biomarker research hinges on the staging of communicative interactions that encourage subjects to produce speech. These interactions, and the subsequent transformations that research personnel carry subjects' speech through, are the source of unanticipated lapses in and additions to the research pipeline. In these instances, researchers at once construct vocal biomarkers as technoscientific objects while also opening up "a little universe of indeterminate and far from unmotivated subject positions."[12] When the technophilic and technophobic discourses surrounding vocal biomarkers sideline this labor, they occlude the patchwork nature of vocal biomarkers— the fact that they are fabricated rather than found. Additionally, these binary discourses fail to voice the values, meanings, surpluses, and relations that researchers and research subjects generate alongside and from within the technoscientific task at hand, all of which remain ancillary or sometimes even antagonistic to the process of dataset creation.

My analysis proceeds in an ambivalent register, neither downplaying the violent and exploitative nature of vocal biomarker research, nor glamorizing transductive labor, nor discounting the small evasions of capture generated through this category of work. In what follows, I expand on my definition of transductive labor, demonstrating that, despite its marginalized status, it is suffused with the potential for unanticipated overflows. I then provide two examples of transductive labor's unheard excesses in context. The first illustrates the unauthorized practices of care that researchers and research subjects forge along the research and development pipeline. The second illustrates how some subjects' decision to remain silent introduces uncontrollable statistical error into the dataset. Listening to machine listening against the grain of its more dichotomous depictions, I follow the directives of Black feminist sound studies and disability studies scholars to attune our senses to signs of alterity operating on quieter, "lower frequency" levels, focusing less on the technological specifications of algorithmic voice analysis and more on the relations that they create, rupture, or re-articulate.[13]

Glitchy Signals: From Transduction to Transductive Labor

Feminist and critical race science and technology studies scholars have shown that modern institutions of power use technologies such as full-body scanners and automated facial recognition systems to compress gender, race, sexuality, disability, and citizenship into static, visual signifiers through which to interpret and manage deviance.[14] Technologies for surveilling sound, like algorithmic visualization technologies, similarly naturalize complex, sociopolitical phenomena into rigid, decontextualizable types. Vocal biomarker technologies reduce mental distress and madness, and the fraught colonial, White supremacist histories subtending dominant definitions of "rationality," into a sign of engrained, audible deviance.[15] Yet examining vocal biomarker research in its day-to-day unfolding ethnographically, at the level of the labor required to sustain it, reveals the incompleteness of this reduction, or what I refer to as the politics of things not heard.

To produce a device that "recognizes" a pathological speech sound, researchers must define how pathological speech *sounds*. They must gather a dataset of audio-recorded speech from human research subjects experiencing mental distress and then use this dataset to construct a standardized threshold for distinguishing "normal" vocalizations from "mentally ill" ones. Achieving this end depends on a series of transformations. Research personnel and human research subjects must transmute the subjects' speech from one form (spoken utterances) into another (the informatic argot of an automated system). I call this transformational work transductive labor. I follow sound studies scholars, media historians, and anthropologists who have taken up "transduction" (from the Latin *traduscere*, to lead across or transfer) as a technical descriptor of technologies that reproduce, transmit, and process sound and as an analytic emphasizing materiality, partiality, and difference. Transductive labor draws even closer attention to the workers who tend to and enable the functionality of transduction-dependent media, along with the quiet instances of unruliness associated with this work.

Technologically speaking, transduction occurs when a person makes a telephone call on a landline and the sound of their voice is transformed into electrical impulses, sent across wires, and transferred back into sound in the receiver of their interlocutor's phone. Whereas transducers are devices that "turn sound into something else and that something else back into sound," transduction refers to the more general process of converting "a signal from one medium to another."[16] Many technoscientific endeavors involving audio processing depend on transduction. For example, creating a spectrogram, or a visualization of a sound's frequency over time, involves converting sound from audible, physical energy into a graphic representation. Fleshly bodies, too, transduce. Studies of the physiological conversions through which the bodies of hearing and d/Deaf individuals process sound marked a key pivot point in the development of the telephone.[17] As Mara Mills writes, Helen Keller, a deafblind woman from whom cybernetician Norbert Wiener drew (audist) inspiration, described herself as a kind of transducer: "Like the telephone and other electro-acoustic technologies, she was a partial translator of sensuous phenomena, themselves connected through the universal language of physical oscillation."[18]

Conceptually, transduction throws a mediating wrench into technodeterminist depictions of information and communication technologies as producers of frictionless connectivity.[19] Focusing on transducers brings into earshot a web of social and technical infrastructures that prop up the conversion process and interrupt it when they break down. That is, rather than facilitating a seamless translation of sound from one medium to another, because of transduction's unavoidably material and thus unstable nature, it leads to misalignments and haphazard correspondences between original input and transduced output, generating what Julian Henriques calls a "surplus."[20] To Henriques, this unaccounted-for excess holds politically potent potential, rendering a transducer into "a device for achieving escape velocity from the world of either/or to the world of either or both," a realm of openness that resists reduction into binary absolutes.[21]

This unruliness trickles through transductive labor. Though economists have used the term to describe "work that orients itself to the care and keeping of life through time," my usage focuses on practices of maintenance and care specific to information and communication technologies.[22] Transductive labor includes the manual and menial work that upholds the experience of these technologies as im-mediating (not mediated), such as a telephone "connecting" callers or an algorithm "recognizing" vocal biomarkers.[23] Akin to Julia Elyachar's notion of "phatic labor," transductive labor is geared toward the functionality of social and technoscientific communicative channels.[24] In the case of vocal biomarker research, this includes the provision and preparation of materials (speech) destined to be transformed into something else, or the initial, sonic inputs on the front end of a transductive chain.

Like transduction, transductive labor peels back the illusion of immediacy, amplifying an array of technicians, stewards, and data custodians who facilitate sonic conversions through repair, rapport, and upkeep. At the same time, like other forms of custodial work that maintains automated technologies, transductive labor is figured as mechanical, unskilled, feminized, racialized, and altogether marginal to the technology development process.[25] The transductive laborers on Klaus's team, two women whose work I expand on below, were a salaried research assistant and an undergraduate psychology intern. Compared with others on the team—primarily men trained in engineering—they spent the most time interacting with research subjects, eliciting and listening to the subjects' emotionally intense narratives. Neither woman had substantive clinical experience, nor did the study's host institution provide them with training or dedicated mental health resources. In the team's weekly meetings, the feedback that both women offered regarding the study's design and its impact on research subjects was often downplayed.

Despite this low valuation, with reference to the multiplicity attributed to transducers, the individuals engaged in transductive labor create glitches along the production, research, and development pipeline. Even while entangled in extractive arrangements of data, bodies, and capital, transductive laborers can etch out escape valves and entryways, making space for subversive subject positions and modes of exchange. These things not heard that undergird the production of algorithmic listening exceed, and thus can interrupt, the captivating potentials of the algorithmic system being built. I turn next to examples of this interruption in situ, focusing on two ethnographic episodes of being otherwise that escaped absorption into the team's algorithmic system.

Clandestine Care

Returning to Klaus's office, let us pan past the frame of the videos he shows me to a room on the opposite side of the wall from the subject and avatar. There, in the videos I had watched, Nasrin, an Iranian American undergraduate intern, and Taylor, a White American research assistant, had observed the subjects via live feed. Throughout several of the interviews, the two women had in fact been manipulating the avatar, clicking through dashboards that they developed by conducting, recording, and coding face-to-face interviews with research subjects. The face-to-face interviews also produced an initial corpus of speech data, which the engineers used to develop a preliminary workflow for training the AI model that they would eventually embed into the VHS system, adding metadata labels to quantify the subjects' vocal features based on a combination of prior vocal biomarker research and the opinion of several clinicians with whom they consulted. As Nasrin and Taylor interacted with the subjects through the avatar, their main operative was to draw out more speech to add to this corpus so that the engineers might further refine their classificatory scheme, iteratively fine-tuning their AI model. Nasrin's dashboard controlled the pacing of the avatar's interview questions, modified from gold-standard assessment questionnaires for PTSD and depression. She also steered the avatar's "back-channel" talk: *mm-hmm*s and *uh-huh*s meant to insinuate that the avatar is engaged in the subject's responses and wants to hear more. Taylor operated the avatar's body, making it nod its head, smile, and lean forward with interest.

In the context of their team's study, this Wizard of Oz (WoZ, or "wizarding") experimental setup served two purposes (fig. 10.1).[26] It allowed Nasrin and Taylor to choreograph an interaction that encouraged the subject to speak continuously, producing the acoustic data that the engineers used to train their AI model. It also allowed Nasrin and Taylor to identify which interactional flourishes provoked a flood of vocalizations from the subject and thus should be folded into the automated version of the avatar in the final prototype, ensuring that the avatar prompts the user to produce a steady stream of data for its soft-

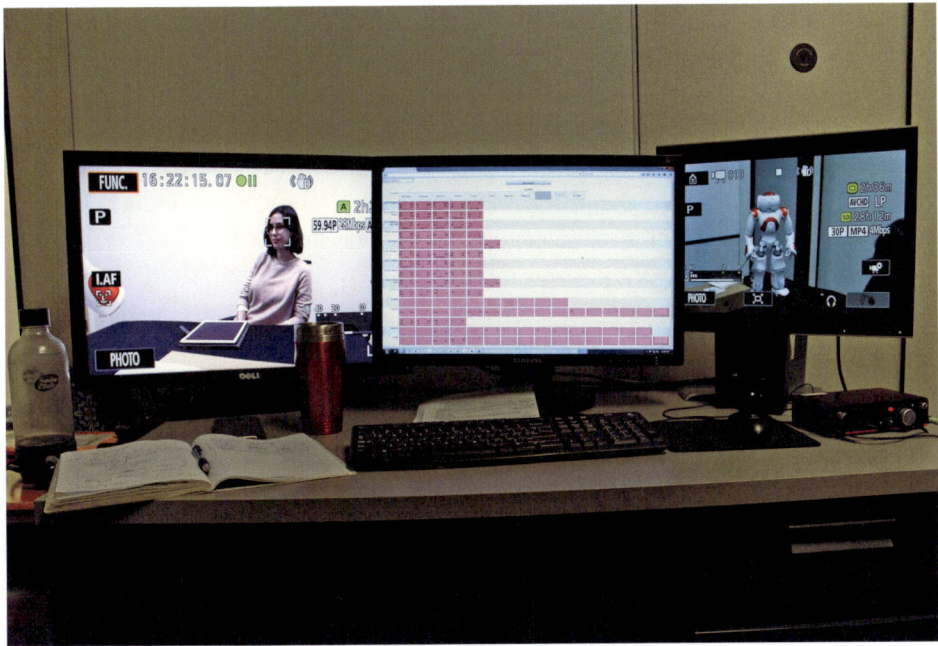

FIGURE 10.1. The author (*left screen*) interacting with a robot (*right screen*) in a WoZ setup, shown from the perspective of a wizarding researcher. The researcher remotely operates the robot's speech and movements from a separate room by selecting options on the dashboard (*center screen*).

ware and the AI model to analyze. The team hoped that, once fully developed, the avatar will standardize the interview/data-extraction process, removing the need for an exhaustion-prone, auditorily imprecise human interviewer.[27]

Although Nasrin and Taylor were physically cloistered from the research subjects for most of their tenure on the study, they revealed to me in an interview that their work had engendered an intense affective investment in the subjects' well-being. Recruited from neighborhoods surrounding the university, throughout the study's various phases, many of the subjects delved into their experiences as veterans or as unemployed or unhoused individuals living through poverty, addiction, and racialized and gendered violence. In the first third of the study, when Nasrin and Taylor conducted face-to-face interviews, the women could offer a more expansive array of reactions and expressions of intersubjective engagement in response to a subject's answers. For the remainder of the study, they only interacted with the subjects face-to-face in order to explain the study's premise and that their interviews with the avatar would be recorded, and to debrief the subjects once the interview ended. Taylor lamented that in these later stages, the subjects lost the experience of "*seeing* you in that way" they had in the face-to-face phase. As Taylor and Nasrin receded from the subjects' view and merged together through the avatar, a somewhat rigid mechanism of data extraction and abstraction, the subjects could no longer witness the weight of their stories move through the women. The women still reacted with each other in the wizarding room, sharply drawing in and holding their breath at tense moments or weeping at cathartic ones. During the study's final phase, despite the fact that the system was technically automated and could run without wizarding, for legal liability reasons and due to the system's inability to process semantic meaning, they continued to monitor the subjects, vigilant for mentions of suicidal ideation while maintaining the illusion that the subjects were "alone with the computer," as Nasrin explained. When debriefing the subjects, they could not reveal that they had observed the interaction, even in the event of disturbing missteps, such as the time the avatar responded, "That's great!" to a subject discussing their spouse's death. Taylor clarified that, after all, the purpose of the system was to passively listen, drawing out data rather than intervening or providing treatment. Nor were she or Nasrin qualified or authorized by the university to administer biomedically sanctioned care. Detachment was a design feature of the study and its end product.

Nevertheless, in the midst of their mandated distance, Nasrin and Taylor engaged in clandestine forms of what Hiʻilei Julia Kawehipuaakahaopulani Hobart and Tamara Kneese call radical care: "strategies for enduring precarious worlds" that presented "an otherwise" at the edges of the study's restrictive modes of intersubjectivity.[28] After the interview and debriefing had finished, Nasrin and Taylor would sit with the subject, not to intercede on or educe what they had overheard but to simply remain in each other's mutually acknowledged presence, sometimes in silence. For as long as the subjects wanted, they held together a brief moment of pause between being enrolled in the study and returning to their lives, an interaction that would never be absorbed into the system as data. Nasrin and Taylor's role in the study as data elicitors put them in intimate proximity with the subjects and the details of their lives. While at times deeply painful and disturbing, continually listening to the subjects, both as themselves and clandestinely through the avatar, laid the groundwork for a sense of responsibility toward the subjects' well-being, catalyzing Nasrin and Taylor to engage in practices of distant yet attentive watchfulness and care-full, collaboratively sustained silence. These forms of informal and subtle repair work were orthogonal to the production of computationally tractable voice data.

Research subjects were paid a stipend of around one hundred dollars in exchange for their voice data, which were of indispensable scientific value to teams like Klaus's. Without the subjects' voices, researchers cannot lay claim to the existence of vocal biomarkers or their system's ability to identify them. Though there were subjects who departed from the study feeling exploited, research subjecthood was not an extractive experience for all. Some told Nasrin and Taylor that being listened to by the avatar was therapeutic. Others took pleasure in the encounter, telling the avatar jokes or pretending to seduce it, transforming the study into a projective screen of fantasy and desire. Still others engaged in a practice I observed across the three sites in which I conducted my fieldwork: saying nothing.

Each of the teams with whom I conducted fieldwork deployed different strategies for eliciting voice data from the subjects, from prompting them to produce a single, elongated vowel sound during a brain scan to undergoing symptom assessments over the phone with a clinical worker to Nasrin and Taylor's avatar-mediated interviews. In turn, some subjects met these strategies with countervailing acts of reticence. While assisting with brain scans at one field site, I would witness subjects repeat the wrong vowel sound or say the sound for too short of a duration, despite the numerous and gentle corrections of research personnel. Researchers understood that subjects' participation was voluntary and revocable, according to university ethical review board protocols for conducting human subject research. Reticent subjects likewise seemed to acknowledge that the definition of "participation" was up to their own, pliable elaboration. In a video I reviewed with Klaus, for instance, one subject either ignored the avatar's questions or offered only single-word answers. When the avatar posed one of its grimmer inquiries ("What's a memory you wish you could erase from your mind?") the subject spat back, "Don't know."

Aggregated into a training dataset, these sparse responses could do delicate damage. Because the teams were interested in mathematically modeling the affective texture of the subjects' voices, brevity (the fleeting presence of speech sounds) or silence (the absence of speech sounds altogether) left them with little or no material to analyze. For this reason, subjects who opted not to speak introduced noise—meaningless waste—into the study. At another field site, when helping researchers add metadata labels to subjects' audio, we were instructed to discard segments with few vocalizations because they added little to no mathematical value to the corpus. Whether shrinking the volume of usable data or expanding null values in the dataset, these unusable data diluted the statistical power of the relationship between the subjects' diagnoses, the state of their bodies and minds, and their voices, watering down the scope of the claims that researchers could make about their algorithm's ability to identify vocal biomarkers beyond the lab.

In fraying the downstream functionality of the automated system, silent subjects also evaded enrollment into the final product and, thus, complicity in fueling the surveillance of subsequent research subjects or users. Whether purposeful or not, their noncompliant quiet introduced pockets of things not heard into the study: absences that disrupted the penetrative depth of the algorithm the study was designed to produce. Subjects could exit the study with the research stipend while having kept their voices to themselves. Technically, they had done their due, transductive diligence, offering data to be transduced in exchange for the stipend—just very bad data. In the process, they scrambled the logic of capitalist exchange central to the study and the enterprise of human subject research writ large, voiding the scientific value of their data while gaining access to a potentially significant amount of money.[29]

Listening to Transductive Labor

As transductive laborers, Nasrin and Taylor set the conditions of possibility for research subjects' speech to be transformed into something else. They facilitated the extraction of data from the voice by guiding conversational interactions with research subjects through the head nods and back-channel chatter of the avatar, itself the product of their face-to-face conversations with subjects. In turn, by responding to the interview questions (or not) research subjects provided the preliminary audio data from which the engineers assembled a voice-recognition algorithm. Researchers' and research subjects' contributions and disruptions to the overall goal of the study remain unheard in multiple ways. The influence they hold over the study's statistical efficacy remains absent from the majority of research articles and public discourse. Moreover, the connections researchers and research subjects forged or the silences they sustained remain absent from the dataset the study is geared toward producing, blots of opacity that cannot be operationalized into a system designed to naturalize associations between sound and disability.

The gaps that workers produce as they move speech across media are more pinpricks than puncture wounds, operating in subtle and even contradictory registers. In this way, attending to transductive labor and its things not heard parallels a method that Tina Campt calls "listening to images." Campt approaches listening not as a practice of discerning audible data but as "an inherently embodied modality constituted by vibration and contact" between a sound's source and its perceiver.[30] This mode of listening runs against the masculinist and, as Dylan Robinson illustrates, settler colonial impulses of technoscientific listening, which is geared toward the extraction of discretely knowable essences, universal truths, and use-values.[31] In contrast, Campt pursues a haptic and relational listening tactic as part of a larger project to expand the visual archive of the African diaspora to include materials that might otherwise be passed over as unremarkable artifacts of the state's attempt to objectify and catalogue Black subjects, such as convict photographs from a Cape Town, South Africa, prison in the early twentieth century. She suggests that we "look beyond what we see" in these materials and instead listen to them, tuning in to the less legible signs of dissent.[32] She compares these "quiet affects" to sound that vibrates at such a low frequency that it is experienced as more of a rumble in the chest than a reverberation in the ears.[33] Listening to quiet affects in this embodied way calls into question the relational interplay between viewer and photographic object while acknowledging that defiance can exist alongside and within the strictures of dispossession and disaffection.[34]

Like Campt's strategy of listening to images, affixing analytic attention to the transductive labor that fuels automated voice analysis technologies turns up the volume on quiet affects—not heard in dominant discourses—that exceed binaries of either/or, such as victim/perpetrator, pure/sullied, or signal/noise. Moreover, it draws our focus to the relations that gather around this category of work. The care that Nasrin and Taylor can offer research subjects is, technically, not care, at least not according to the professional, legally binding, and institutionally ratified standards of the American mental health-care system. Providing neither treatment nor cure, the women instead open up a pocket of stasis and recognition that serves no immediate purpose to the study. At the same time, their wizarding work fomented a sense of responsibility for the subjects that exceeds the protocol of mandated detachment, a "technoaffective attunement" toward the subjects' post-interview state, honed over time through the increasingly hidden proximity to subjects that their transductive labor afforded them.[35] Subjects who experience their encounters with the avatar, or Nasrin and Taylor as avatar, also exceed this mandated detachment; no one can prevent or police its potentially therapeutic or fantastical effects. Nor can anyone on the research team resolve the paradox of subjects who participate through forms of nonparticipation, producing noise for the researchers and gain for themselves.

Listening to transductive labor therefore also makes room for one of the key insights of critical disability studies: that individuals who depend on noninnocent technologies for survival and sustenance can engage in "unfaithful" relationships with technoscience, which neither embrace nor fully reject technofixes and instead push us to take "responsibility for the social relations of science and technology."[36] Research subjects who offer up their stories and speech in exchange for what can be a impactful source of income are more than "mute subjects of governmentality," even while their role in the study enables (or interrupts) the production of harmful surveillance technologies that figure mental distress into biological abhorrence deserving of calculated control.[37] Critical readings of vocal biomarker technologies that neither praise them nor solely critique them for their privacy-violating capacities turn a careful ear to the impartial, strategic relations formed around or in spite of transductive labor. They push us to ask after the social and political arrangements that have made the automation of psychiatric listening—and the continuous, auditory surveillance of people with psychosocial disabilities—seem necessary and desirable in the first place.

NOTES

1. First names referenced without surnames are pseudonyms.

2. Amazon Web Services, "Your Health Speaks: Vocal Biomarkers and the Potential for Direct Measures of Health from Voice," video presentation by Jim Harper, July 11, 2019, YouTube video, 21:01, www.youtube.com/watch?v=4HG6oDBJXfA. See Cummins et al., "Review of Depression and Suicide Risk Assessment Using Speech Analysis."

3. David Adam, "Machines Can Spot Mental Health Issues—If You Hand over Your Personal Data," MIT Technology Review, August 13, 2020, www.technologyreview.com /2020/08/13/1006573/digital-psychiatry-phenotyping-schizophrenia-bipolar-privacy.

4. John Nosta, "Voice as the New Vital Sign," Forbes, October 23, 2018, www.forbes .com/sites/johnnosta/2018/10/23/voice-as-the-new-vital-sign.

5. Quote in Elizabeth Svoboda, "AI Can Detect Signs of Depression. Should We Let It?," Boston Globe, December 8, 2022, www.bostonglobe.com/2022/12/08/opinion/ai -can-detect-signs-depression-should-we-let-it.

6. Weitzel, "Audializing Migrant Bodies."

7. "Do I Sound Sick?"

8. On the racially hegemonic perceiving subject, see Rosa and Flores, "Unsettling Race and Language," 628. For extended treatment of the Whiteness of the liberal speaking subject, see Weheliye, "'Feenin'"; and Navin Brooks, "Fugitive Listening."

9. Quote in Joseph Turow, "Shhhh, They're Listening—Inside the Coming Voice-Profiling Revolution," Conversation, April 28, 2021, https://theconversation.com/shhhh -theyre-listening-inside-the-coming-voice-profiling-revolution-158921. See also Amy Klobuchar, letter to Alex Azar (secretary of US Department of Health and Human Services), December 11, 2020, US Senate (website), www.klobuchar.senate.gov/public /_cache/files/3/5/35297230-1088-4728-80a6-6a8cd1108a9a/2B62F5B3B4E2A4E04A16 A9D01759FF38.121120healthprivacyletter.pdf.

10. For a critical overview of the racist, sexist, and transphobic nature of two such machine vision data sets, see Alex Hanna, Emily Denton, Razvan Amironesei, Andrew

Smart, and Hilary Nicole, "Lines of Site," *Logic(s) Magazine*, December 20, 2020, https://logicmag.io/commons/lines-of-sight; and Prabhu and Birhane, "Large Image Datasets."

11. Semel, "Listening like a Computer."

12. Inoue, "Word for Word," 218.

13. Campt, *Listening to Images*, 6.

14. See Browne, *Dark Matters*; Browne, "Digital Epidermalization"; Beauchamp, *Going Stealth*; Benjamin, *Race after Technology*; and Thakor, "Deception by Design."

15. For an extended discussion of the relationship between normative definitions of rationality and sanity with Whiteness, see Bruce, *How to Go Mad without Losing Your Mind*; and Pickens, *Black Madness :: Mad Blackness*.

16. Sterne, *Audible Past*, 23; Roosth, "Screaming Yeast," 338.

17. Mills, "On Disability and Cybernetics."

18. Mills, "On Disability and Cybernetics," 75.

19. Helmreich, "Anthropologist Underwater"; Helmreich, "Transduction"; Hsieh, "Piano Transductions."

20. Henriques, *Sonic Bodies*, 409. See also Silverstein, "Translation, Transduction, Transformation."

21. Henriques, *Sonic Bodies*, 469.

22. Goudzwaard and de Lange, *Beyond Poverty and Affluence*, 57.

23. Another salient example of transductive labor can be found trailing behind commercial voice analysis technologies, like Amazon's Echo, a voice-activated home automation device with an anthropomorphized digital assistant, or user interface, called Alexa. Propping up the device are thousands of contracted and full-time workers in India, Costa Rica, and Romania, who parse snippets of user interactions with the device in order to improve its functionality. This job can expose workers to disturbing soundscapes, such as recordings of domestic violence and sexual assault. As Thao Phan underscores, the White aestheticization of the Echo's digital assistant via its "accentless" feminized voice glosses over both the figure of the colonial domestic servant that the device mimics and the present-day global networks of racialized labor from which the device benefits. See Phan, "Amazon Echo and the Aesthetics of Whiteness," and Matt Day, Giles Turner, and Natalia Drozdiak, "Thousands of Amazon Workers Listen to Alexa Users' Conversations," *Time*, April 11, 2019, https://time.com/5568815/amazon-workers-listen-to-alexa.

24. Elyachar, "Phatic Labor, Infrastructure, and the Question of Empowerment in Cairo." See also Jakobson, "Closing Statements."

25. See Atanasoski and Vora, *Surrogate Humanity*.

26. See Steinfeld, Odest, and Scassellati, "Oz of Wizard."

27. Elsewhere, I describe the team's use of gender (feminization) and race (Whiteness) to configure the avatar into an alluring yet docile listening subject. Semel, "Speech, Signal, Symptom."

28. Hobart and Kneese, "Radical Care," 2–3.

29. See Petryna, *When Experiments Travel*.

30. Campt, *Listening to Images*, 40.

31. D. Robinson, *Hungry Listening*. See also Eidsheim, *Sensing Sound*.

32. Campt, *Listening to Images*, 9.

33. Campt, *Listening to Images*, 3.

34. Campt, *Listening to Images*, 3–4.

35. On "technoaffective attunement," see Amrute, "Of Techno-ethics and Techno-affects," 70.

36. Kafer, *Feminist, Queer, Crip*, 123–25.

37. Quote in Campt, *Listening to Images*, 9.

Street Smarts   Katherine Bennett

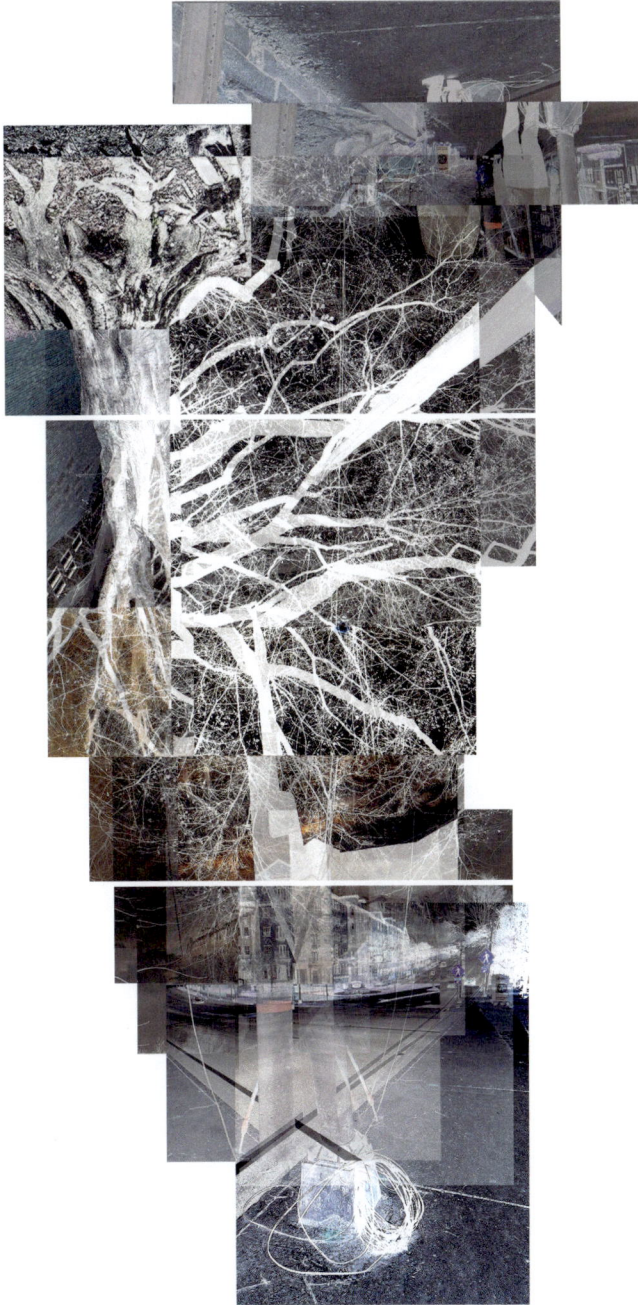

FIGURE INTER4.1. *Street Smarts* by Katherine Bennett

Street Smarts maps quotients of intelligence not normally recognized by proponents of big data and the big real estate, among other things, Smart City technology serves. This digital-analog mesh of services bounds the lives of urbanites it reverse-engineers through computational surveillance systems deemed intelligent. But the trees, people, and myriad other beings watched and measured have the street smarts to push back, literally pushing up the pavement, grinding it down, bending it as it bends them.

Street Smarts represents the city street as a place of change rather than pixels of data and territory. The project's letter-sized, variably folded, spliced, and annotated paper maps interrupt alignments of property and service. The maps locate elder oak trees living along avenues and boulevards, navigating a mineral forest of cars, cables, and tunnels. The oaks know this terrain better than its oldest human residents and highest resolution cameras. The oaks have shaped it for centuries. Their branches trace the street's history, its architectures, fires, and storms. Its rising and falling social status. Their roots penetrate its surfaces, moving air, water, earth, and information underground. The oaks negotiate currents that people sense by feeling as much as seeing—concrete waves of uplifted sidewalks, cool breaths on stagnant afternoons, crumpled demolition notices and soggy stuffed animals, hacked wires, unprescribed drugs, untracked bets, a tangle of shadows. These creative intelligences remap living histories and changing contours of the street.

Interlude

Smart Forests

The integration of digital technologies in nature under the label of smart forests conjures the ideal of a techno-utopic future: where the temperature, moisture, and nutrient content in the soil and air are monitored continuously, where data about logging or warming are made available to support "smart" decision-making, where scarce resources such as water are managed efficiently, and where ultimately the negative environmental impacts are sensed and defused. Yet, we might ask, is the warming of the planet, deterioration of the land, deforestation, and death of species due to our inability to map and measure? Or are they of a piece with extractive economies and silencing of alternative politics?

The idea that digital technologies hold the key to environmental sustainability is contradictory given the devastating environmental impacts of the mining of precious minerals and energy-intensive manufacturing processes required to produce these technologies and the immense amount of e-waste generated at the end of their life cycle. Smart forest initiatives are proliferating nonetheless. For years, Moore's law—Intel cofounder Gordon Moore's theory that the number of transistors on a microprocessor chip would double every two years—has been the driving economic and creative engine for the semiconductor industry. Yet as this principle reaches its limits, the industry needs to find new ways to continue the pace of production and integration. Here, the Internet of Things and its associated platforms and applications such as smart cities or smart forests offer a way forward: we may not be able to double the speed. But we could instead augment everything with computing—or let technology creep into everything—be it trees in forests or traffic lights in cities.

Smart forest initiatives differ in aim and scope as well as in the kinds of technologies that they employ. Yet a quick survey reveals that they are broadly aimed at "banking nature" through a centralized system driven by the ideals of total monitoring and management.[1] The US Department of Agriculture's Forest Service, for example, centers its smart forest initiatives around the deployment of nearby communities to collect and transmit essential environmental measurements wirelessly from strategically distributed sites across different geographic, climatic, and vegetation gradients.[2] Real-time access to sensor data is provided through a website, and visualization and outreach tools are employed to engage researchers, resource managers, educators, and the public with the collected data. The initiative promises the development and implementation of cyber infrastructure to improve environmental monitoring at these sites.

There is much that can be said about the integration of sensor technologies into nature. For example, one might question how they are being deployed with little or no attention to their social, political, and human consequences; how they undermine local and Indigenous modes of knowing and acting with nature; how they potentially shift the loci of power and decision-making away from nearby communities to centralized governmental or private actors; or how they extend and expand economies of extraction.[3]

David Rojas situates smart forest initiatives in a geopolitical perspective and maps their imbrication in settler "developmental" and capitalist operations. Brazil's extremist right-wing agro-industrial initiatives under Jair Bolsonaro's presidency (2019–22) resulted in massive deforestation and violence against Indigenous peoples in Amazonia. Against this backdrop, neoliberal "climate-smart initiatives" have been positioned as a solution to the devastation. However, Rojas argues that fascist and neoliberal approaches cannot be the only two approaches to deforestation, because both ultimately fail due to their underlying profit-oriented ethos. Thus, even "climate-smart" policies have been "promoted as a way to open a future of endless economic expansion" under the pretense of forest preservation. Fundamentally, Rojas argues, they are still based on the premise that "whatever challenges we face in the present, we can transcend them through an aggressive modality of labor in/ of time, wherein industrial tools allow humans to transcend Amazonia's past and present and, from this elevated position, to take the region to its historical destiny."[4] Put otherwise, "relying on capitalist, profit-oriented socio-ecological assemblages to create new futures [will not work]. Their short-term, profit-oriented focus fails to grapple with the nonproductive, deep temporality of processes such as decomposition and soil formation, which are advanced by myriad nonhuman beings whose behaviors and relationships are recalcitrant to human mastery."[5]

Smart forest initiatives insist on a positivist outlook—one that frames nature as a material resource that can and should be put to optimal use through sensing mechanisms (fig. Inter4.2) and algorithmic decision-making. As Jennifer Gabrys puts it, "Trees become carbon sinks, low-lying vegetation acts as flood defenses, shrubs and vines take up air pollution, and mass planting mitigates urban heat island effects."[6] Nature is positioned as a scarce resource that must be attended to not in its own right but rather to ensure development in a capitalist and expansive sense. For example, sensors may be deployed to monitor soil temperature and moisture or track deforestation. Yet the readings are inherently partial, limited to a particular moment in time within a constantly transforming landscape. Thus, as Rojas argues in relation to Amazonia, the scientists whose research is meant to influence policy must constantly confront the limits of what they are able to know through the sensor data. Rojas quotes Caio, one of the scientists working with climate-agriculture data from Amazonia, about his experience with climate-smart initiatives:

"You get your results," Caio said, alluding to research on issues such as the relation between precipitation and tree behaviors in Amazonia, "and then you study more and then you get another [result]." "So then [in this data-gathering process]," he continued, "you bring down truths. And construct new truths. So the question is: when will [scientific knowledge] be ready to be used [in policy efforts]? Never!" Eschewing recourse to final "truths," Caio's disjointed labor in/of time rejected ambitions to craft total representations of socio-environmental orders to come and instead focused on tracing up close a wide range of irresolvable contradictions. *This meant tracking past occurrences as they reverberated in the*

FIGURE INTER4.2. Typical image of smart forest technologies

present and the future and being attentive to nonhuman creativity that remains recalcitrant to human mastery [emphasis added].[7]

As a method, creep can help us heed this call to be attentive to *nonhuman creativity that remains recalcitrant to human mastery*. An initial look at the integration of monitoring technologies in nature as creepy and creeping might surface issues akin to those addressed in the preceding chapters: the dominance of neoliberal ideas and ideals; the accumulation of technologies that depend on a lot of energy alongside the often-invisible labor and maintenance, and the reinforcement of binary categorizations such as useful and idle, human and nonhuman. Here, however, we center questions related to what constitutes knowledge and whose knowledge counts. The gradual, slow movement that is associated with creep expands our conception of nonhuman intelligences and networks, pushing back on technological acceleration in ways that can highlight non-capitalist orientations of grassroots social change. Following the slowness of plant growth and soil movement shifts the meaning of "smart" in "smart forests" from a ubicomp dream of proliferating smart sensors. A more capacious approach to intelligence, which includes human and nonhuman senses, echoes and centers non-human and more-than-human ways of knowing and being in relationship.

We feel and breathe forests and are otherwise touched by their presence. We may not see forests, however, except from a distance. Monitoring forests with digital technologies detaches us from their felt qualities and further distances us from the kinds of intellectual, embodied, and emotional understandings that are crucial to protecting them. The monitoring's attention to relationality in time and space is also limited. Not only that, what digital technologies allow us to see of forests is deeply skewed. Suppose we are interested in capturing the movements of tree leaves in the wind within a day. We might install a series of video cameras to capture these movements from multiple angles; we could add a chip to each leaf and capture its position. This would be consistent with smart forests initiatives in that they aim for a precise all-encompassing view (the God trick!). It is a view that is about predictions and control over the object being monitored. Alternatively, we might do as British artist Tim Knowles does: adding a pen to the tip of the branch, we might produce drawings that capture the movements of branches (figs. Inter 4.3 and Inter 4.4). In doing so, we would begin to see different patterns, ones that we could not glean from monitoring sensors, but are noteworthy and beautiful nevertheless.

At first glance, Knowles's approach to capturing tree movements may appear strange. But it is in no way stranger than when we say we try to understand forests and trees through sensor data. Both the sensors and the tree branch give us traces and projections of phenomena limited by the instruments' capacity and the human aims that put them to work.[8] Both are inherently biased in the data and images that they produce. What distinguishes Knowles's tree drawings is their surrender to the tree for guidance, and, perhaps even more importantly, surrender to accomodate always-already partial and situated perspectives. The tree determines the placement of the easel. The tree renders the drawing, which tells a story of a particular branch's dance in the wind. The temporality of the drawing, too, is highly contingent on the tree and the wind.

Knowles's artistic intervention shows one pathway into a multispecies and multitemporal strategy. This aligns with Rojas's argument that the point should not be "to criticize CSA [climate-smart agriculture] from an elevated, heroic position that would allow us to pass judgment on those who failed before us, right the wrongs of the past, and trace a brave, new path ahead. The argument, rather, is for taking into serious ethnographic consideration multi-species and multi-temporal stories, ideas, and practices that are in excess of the neoliberal and fascist times that seem to contain them."[9] Creeping, in this sense, may mark the kinds of temporalities of slowness that are also in excess of neoliberal and fascist demands.

Relatedly, as persistent movement and growth, creep may be the only way of acting available to some groups or species, the only chance at life. After all,

FIGURE INTER4.3. *Tree Drawing Hawthorn on Easel* by Tim Knowles

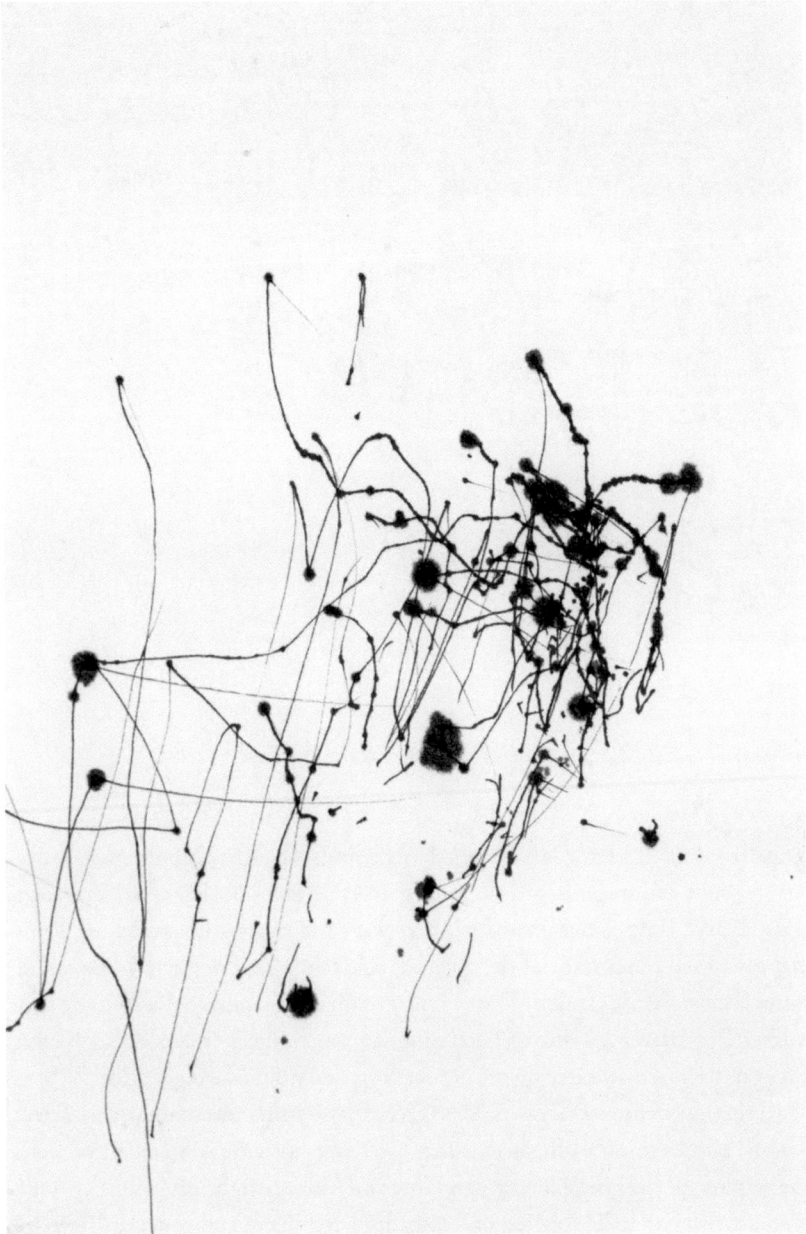

FIGURE INTER4.4. *Tree Drawing Hawthorn on Easel* by Tim Knowles

FIGURE INTER4.5. Katherine Bennett's *Street Smarts* process

"grass [too] has a strategy that works!" as feminist theorist Deboleena Roy observes about the stolonic growth in a species of everyday backyard Bermuda grass: "Over time, I became captivated with the outwardly stretching veins that ran along the surface of the ground, constantly reaching out, in search of connections, feeling around. This is how I realized that grass has a strategy that works. This strategy is one of becoming, and as [Gilles] Deleuze and [Félix] Guattari write, this strategy works at making a communicating world."[10]

Katherine Bennett's artwork that accompanies this interlude, titled *Street Smarts*, pushes this idea further, as she, too, looks into the strategic movement and wisdom of trees (figs. Inter4.5 and Inter4.6). In a thought-provoking critique of smart city initiatives focused on centralized command and control, driven by efficiency and policing objectives that align with technocratic sensibilities, Bennett boldly poses the question: "Could I map the surface conditions of trees while reflecting their subsurface movements to intelligently warp the avenue?"

Bennett traces tree roots' movement as they creep outward and downward, purposefully pushing against concrete. Rethinking smart initiative fixations on

FIGURE INTER4.6. Katherine Bennett's *Street Smarts* process

measurement and control, she turns her attention to the neglected intelligences that surround us. She writes, "In this project, I'm looking for ways to map quotients of intelligence not considered in digital, Smart Corridor technology." "Smart" technology maps city streets as sites of "Internet of Things (IoT) deployment, data collection/analytics, autonomous vehicles, and [unspecified] partnerships." The City of Atlanta rolled out such smart-city technology as a "public demonstration and 'living lab' along North Avenue, a now-central arterial street that crosses the city's Midtown. Historically, though, the avenue marked the city's northern limits in the mid-19th century time of slavery, cotton, and paper."[11] Bennett's work is a meditation, to use Alexis Pauline Gumbs's words, on cross-species mentorships, which "open up space for wondering together and asking questions towards a depth of engagement that is still emerging."[12] For Bennett, unearthing the intelligences of the roots is of a piece with unearthing the very local histories that shape and reshape the built environment. This history encompasses a violent past, marked by exploitation of land and slavery, that continues to persistently surface in structural racism and other forms of economic and social marginalization.

Jane Tingley's digital art installation *Foresta-Inclusive* presents us with another evocative alternative to the smart forest ethos. Using the exact same technologies as those that made up smart forest initiatives, she draws attention to the life and livelihood of trees. Her aim is to foster and understand our human

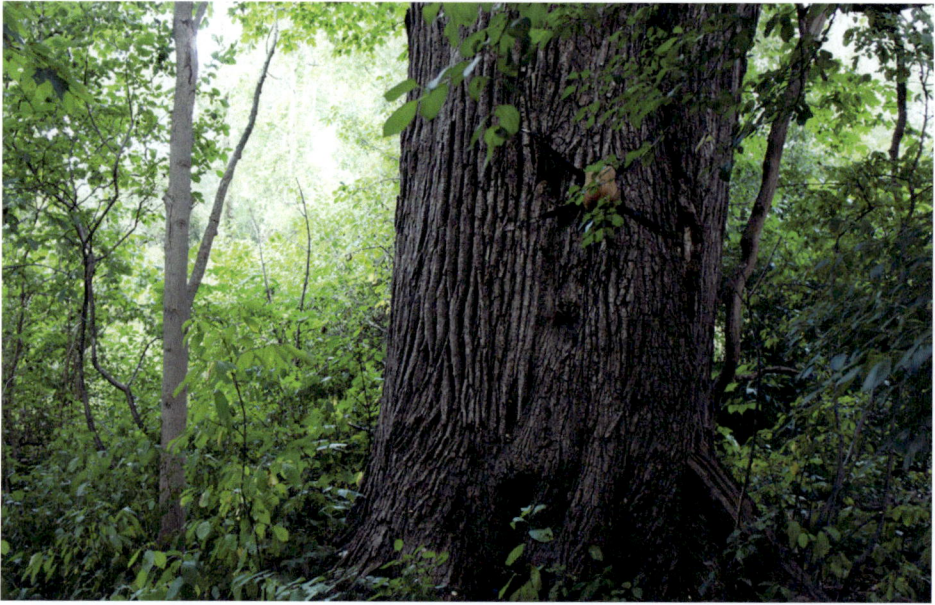

FIGURE INTER 4.7. From the digital art installation *Foresta-Inclusive*, by Jane Tingley

selves and our relations with more-than-human worlds. She uses sensors and visualizations to draw out the complexity and livelihood of nature. Comparing the sensors she uses for her art installations with those typically used in smart forest initiatives reveals a contrast. Hers is suggestive of a collaborative approach, reflected in the form of the sensor box hugging the tree branch (fig. Inter 4.7). We may compare that with images from smart forest initiatives, such as the one in figure Inter 4.2, which depict sensors as wired into nature, by contrast, yet stay alien to it. This contrast serves as an analogy for how the latter pays little attention to integrative and cooperative possibilities. Tingley asks, "What does it mean to be alive and have agency?; How can we re-train ourselves to slow down and listen to voices that have been marginalized for millennia?; What does it mean to be in dialogue with something that does not share the same language nor temporal reality?; and once we acknowledge the 'aliveness' of something, what are the ethical implications of that recognition?"[13] These questions permeate the way she seeks to relate with, to, and through nature.

The imagery that results from this engagement with nature (e.g., fig. Inter 4.8) showcases a feminist orientation to technology, especially in its resonances with a deeper emotional and intellectual commitment to multi-species embodied knowledges. It is an engagement that is inclusive of all our senses,

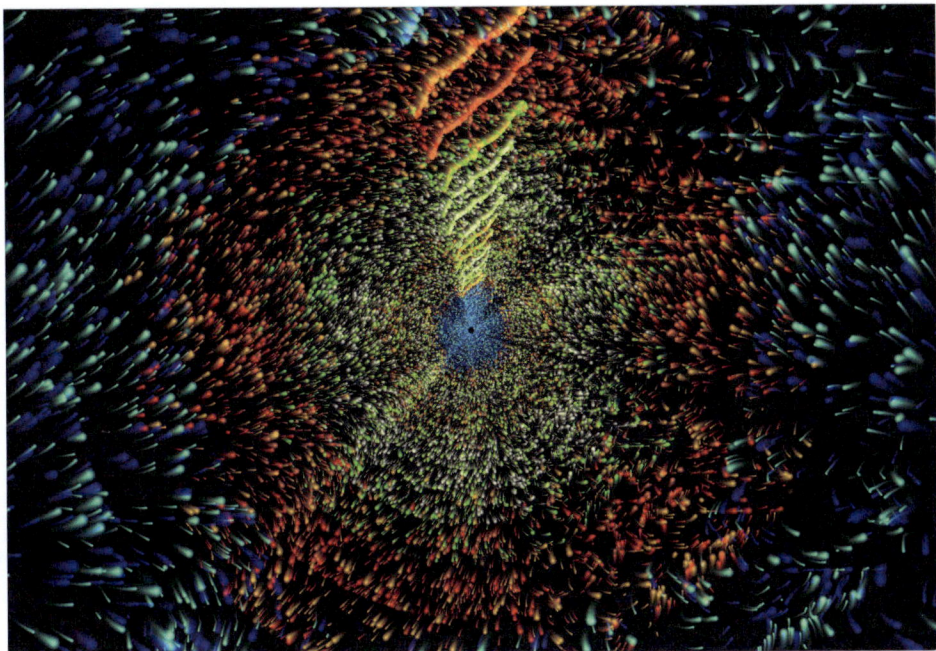

FIGURE INTER4.8. From the digital art installation *Foresta-Inclusive*, by Jane Tingley

where "feeling for the organism" serves as a starting point, as a strategy and methodology, and as an end goal and purpose.[14]

The artworks featured in this interlude, and especially their foregrounding of knowledge that exceeds Western rationalist approaches, orient us toward what Indigenous scholars have theorized as a way of the world that is in tune with the knowledge and wisdom of the more-than-human, and that includes a deep understanding of the interconnectedness of all living beings and environments. In an article evocatively coauthored with Bawaka Country, a Yolŋu Indigenous Homeland in northern Australia, the authors invite readers to dig for *ganguri* (yams) at and with the Bawaka land, and in the process gain insights about "our co-becomings in a relational world."[15] They call attention to what the land knows in order to frame digging as a spatial practice *and* a mode of knowledge production. "The actions and ways of knowing of each party—the ganguri, the sand, the humans (Yolŋu and non-Yolŋu, writer and reader)—are richly contributing factors. Thus Yolŋu co-becoming underpins an engagement with the more-than-human world which nurtures a vitalist sense of coexistence, emplacement and knowledge. This represents a marked shift from dominant, ocular-centric western perspectives of place in which visuality and a certain kind of knowledge are conflated, so that a (human) onlooker can understand and appropriately value a place/space through the act of viewing or writing it from a suitable distance." Reframing knowledge in a way that decenters the visual not only challenges the "hierarchy of knowledge" and settler colonial conceptions of place, but also opens possibilities for expanding our conceptions of collectivity and action. "Bawaka as a co-constituted and constantly emerging space/place prompts us to attend to the more-than-human, and indeed more-than-human-centric aspects of both relationality and ethics."[16]

Kyle Powys Whyte's account of how such ethics and politics can be seen in the grassroots action of Indigenous women offers another powerful example. He begins by highlighting climate change and its associated disruptions, such as sea level rise and shifts in the ranges of species. He points out how such changes disrupt the continuance of the systems of responsibilities for some Indigenous women whose communities live in close connection with the earth. The impacts of climate change on the Great Lakes region in North America, including on "the range, quality and quantity of species like berries," and the degraded quality of water affects the responsibilities and identities of Anishinaabe women, such as for ceremony.[17] These changes expose these women to harm but also showcase the ways that their understanding of their commitments motivates their political action.

Whyte offers the concept of "collective continuance" as "a community's aptitude for being adaptive in ways sufficient for the livelihoods of its members to flourish into the future. . . . According to this view, collective continuance is composed of and oriented around the many relationships within single communities and amid neighboring communities that persons assume based on their culturally framed perceptions of what matters."[18] Such responsibilities shape communal arrangements and modes of action in response to the ongoing context of settler colonialism. Whyte writes: "The flourishing of livelihoods refers to both indigenous conceptions of (1) how to contest colonial hardships, like religious discrimination and disrespect for treaty rights, and (2) how to pursue comprehensive aims at robust living, like building cohesive societies, vibrant cultures, strong subsistence and commercial economies, and peaceful relations with a range of neighbors, from settler towns to nation-states to the United Nations (UN). Given (1) and (2), indigenous collective continuance can be seen as a community's fitness for making adjustments to current or predicted change in ways that contest colonial hardships and embolden comprehensive aims at robust living."[19]

The ethics of robust living connects members to their relatives and ancestors as well as other species such as salmon or natural collectives such as forests or bodies of water. The responsibilities are bound up with these relationships, such as when women grapple with the degradation of water in the sense of grappling with the decline of a close relative. These responsibilities are also variable depending on individual and group standing in relation to others. For example, in the above case, the degraded water is a core existential concern for Anishinaabe women, and their experiences with climate change differ from those of Anishinaabe men and non-Anishinaabe individuals. The women's responses—manifest in part in the Mother Earth Water Walk and the grassroots women's group Akii Kwe, which have led to positive changes in decision-making processes in Canada—are highly specific to their particular circumstance, inclusive of their historical lack of participation in the political processes. "They are forms of collective action based on structures that indigenous women see as furnishing regional and global-scale participation and representation for their culturally inspired concerns and ideas, and that are most appropriate for addressing how their communities are affected by climate change impacts. They are forms of collective action that seek to engender global relations of coexistence on indigenous people's terms."[20]

Communal continuance entails a heightened awareness of the living world, informative of ways that one might move and interact with their surroundings in a symbiotic and harmonious manner. Transcending the all-too-common conviction that we can use, control, and outsmart nature, the idea of communal continuance demands that we trust the intelligence of the more-than-human world. What if we instead orient ourselves to trees and their communal strategies of being and becoming, or follow the movements of creeping plants and their ways of co-existence and codependence?

NOTES

1. Gabrys, "Smart Forests and Data Practices."

2. "'Smart Forests' Digital Environmental Sensors and Telecommunications Take Research to New Levels," https://research.fs.usda.gov/news/highlights/smart-forests-digital-environmental-sensors-and-telecommunications-take-research, accessed July 26, 2024.

3. E.g., see Liboiron, *Pollution Is Colonialism*; and Tsing, *Mushroom at the End of the World*.

4. Rojas, "Disjointed Times in 'Climate-Smart' Amazonia," 327.

5. Rojas, "Disjointed Times in 'Climate-Smart' Amazonia," 322.

6. Gabrys, "Programming Nature as Infrastructure in the Smart Forest City."

7. Rojas, "Dosjointed Times in 'Climate Smart' Amazonia," 332.

8. Loukissas, *All Data Are Local*; Gitelman, *"Raw Data" Is an Oxymoron*.

9. Rojas, "Disjointed Times in 'Climate Smart' Amazonia," 336.

10. D. Roy, *Molecular Feminisms*, 3–7.

11. Katherine Bennett, "Street Smarts," Katherine Bennett (website), 2019, https://kebennett.com/street-smarts.

12. Gumbs, "Undrowned."

13. Jane Tingley, "Foresta Inclusive," https://janetingley.com/foresta-inclusive/, accessed July 26, 2024.

14. Keller, *Feeling for the Organism.*

15. Bawaka Country et al., "Co-becoming Bawaka."

16. Bawaka Country et al., "Co-becoming Bawaka," 463, 458.

17. Whyte, "Indigenous Women, Climate Change Impacts, and Collective Action," 606.

18. Whyte, "Indigenous Women, Climate Change Impacts, and Collective Action," 602.

19. Whyte, "Indigenous Women, Climate Change Impacts, and Collective Action," 602.

20. Whyte, "Indigenous Women, Climate Change Impacts, and Collective Action," 612.

11

Animal-Vegetal-Technology

Creeping Categories

So the true question then is not whether nonhumans can communicate and make meaning; rather, we must ask: When and how did a small group of humans come to believe that other beings, including the majority of their own species, were incapable of articulation and agency? How were they able to establish the idea that nonhumans are mute, and without minds, as the dominant wisdom of the time? —Amitav Ghosh, 2021

Language use has centered man's epistemic arrogance. To speak is to be exceptional, to be human. The command of technology now furthers this civilizational project to become exceptionally human, especially as modernity seems premised on speaking humans flexing technological muscle to advance science and to progress. While humans use technology to further themselves, those beings called animals and plants are curiously deemed bereft of technology and may become technology instead. Amitav Ghosh's words from "Brutes" haunt the hubris behind these demarcations—human, animal, plant—and anchor the hubris as a central problem shaping our present planetary crisis. We continue extolling species distinctions and projecting our anthropocentric exclusions, all the while embedding deep racial, gender, and national divides, among others, into our classificatory schemas. In capitalism, divisions are blurred as necessary for the flow of capital and consumption. Yet ironically, even as divisions are blurred, the politics of capitalism also further divides as a strategy to keep intact social, economic, and planetary rifts. The contemporary global pandemic, rife with technological intervention and development, illuminates this irony of blurring while furthering divisions. Intrigued and troubled by the curious demarcations of living beings as human, vegetal, or animal, demarcations premised on the hubris of human command of language and technology, I find that the COVID-19 era offers an opportunity to question these demarcations, especially with regard to technology as a distinguishing feature among them. In this chapter, I work with the concept of technocreep to explore how technology creeps through the categories of human, animal, and plant in COVID times, even as those categories remain intact. Can we change our understanding of the "human" by employing a creepy methodology?

My framing of our present historical moment as "COVID times" refers to the temporal oscillations that have been characteristic of social life since the declaration of the global pandemic in March 2020. Old and new, past and present, and there and here were jostled and interwoven through viral transmissions that made us question the sacrosanct linearity and boundedness of any time or place: the 1918 influenza pandemic was being used to demonstrate the "here and now" of history; news channels repeatedly broadcast pictures of masked people from 1918 and 1919, an eerie reminder that history was repeating itself; and air travel restrictions arrived after the coronavirus had already hitchhiked and traveled afar and we struggled haplessly just to follow the virus as it traveled through the world, an enormous blow to human hubris and its civilization. "Post"-COVID celebrations were marked by an arrogant naivete that assumed we can start again from where we left off in "pre"-pandemic times and resume the same old patterns of relations. No, these are still COVID times, when human progress is decelerating in "anachronic time," a term coined by Jacques Derrida to describe the unsettling back-and-forth movement that characterizes our present moment. Categories, progress narratives, and temporal linearity are destabilized while

COVID waves through the world, pushing us back and forth and repeatedly accentuating social stratifications, death, and misery.

Mel Y. Chen puts it best with the following words: "COVID-19 both is and isn't the name of a virus. It is many, many things—many histories, many bodies, many politics. It is also the name of differential bodily burdens, differential state resourcing and differential state securitizations under terms that create bifurcations between care and murder."[1]

As Chen writes, COVID-19 is about varied bodies experiencing time in diverse ways. It is impossible to homogenize experiences or bodies negotiating the political economy of the present. Climate change and neoliberalism's intimate relation with COVID times unsettles different bodies and increases the divides between those who live and those who cannot. COVID times are explicitly about border crossing, whether seen through the travels of the virus or the demarcations that it traverses such as public and private, self and other, and animal and human. All the while, it is important to remember, as Elora Halim Chowdhury points out, that "the differential border crossings of the virus have been the consequence of unfettered capitalist development and accumulation."[2]

The role of technology in COVID times is ubiquitous. The salience of technology and its creepy qualities have enveloped life and death along with the global pandemic. Zoom cameras infiltrate our private spaces, and COVID-tracking apps, with their haptic alerts on our phones, have become a security imperative. The method I call technocreep, following the call for this volume, is a method of attunement to the way technology creeps into our everyday life and sensorium and creeps around and through categories even as it strengthens them. Working with technocreep as a methodological approach, this chapter delineates the continuum of humans-animals-plants-technology as part of larger nature-culture questions showcased during COVID times. Technocreep calls for the art of noticing the creepiness of technology as it moves alongside our natural world consisting of animals and plants as they creep and mutate through climate change and shifts in the global political economy. We continue to "create bifurcations," as Chen points out, through a simultaneous politics of blurring boundaries, even in calls to the "we" of solidarity, and accentuating stratifications of class, gender, nations, and species. Intrigued and troubled by the presence and intrusiveness of technologies in COVID times that are breaking divides even as they further them, here I write about the play with categories that are characteristic of our present moment.

The increased social stratifications and strengthening of the categories of human, animal, and plant have intensified the uses of technology, or, in other words, today we use technology to constantly articulate what is human, animal, and plant. I elaborate on this politics in the last section of the chapter, along with the urgency of working with technocreep as methodology. I turn our attention first to a visualization of the subversion of technology, a performance that demonstrates the creepy play of technology in our lives and provides ways to speak against its logic. It also showcases how animal, plant, and human bodies creep outside their projected frames. Artists are often able to showcase the politics of life in a playful manner that remains difficult to achieve otherwise. While technology is also art, and art may deconstruct the uses of technology, the infoldings between art and technology provide intriguing matter for thought and action. The next section engages with the artwork of Mithu Sen to visualize the demarcations we make between technology, animal, plant, and human, as well as to think about technological subversions. Sen's art demonstrates the creepiness of technology, and by centering creep on the boundaries of plant, animal, technology, and human, it also demonstrates technocreep as method.

Mithu Sen: Technological Subversions

Mithu Sen is one of India's most well-known contemporary artists. She has traveled with her exhibits and performances throughout the world. Her solo shows include *I Have Only One Language; It Is Not Mine* (2019) in New York, *Un-MYthU* (2018) in Mumbai, *Border Unseen* (2014) in Michigan, and *Cannibal Lullaby* (2013) in Belgium. Persistently border-crossing frames of representative politics, Sen's artworks are scathing social critiques and playful subversions. Minutely detailed, macabre, personal, and political in the presentation of human and nonhuman bodies, Sen's art includes drawings, performances, poetry, and multimedia exhibits, alongside a spirited play with technology. She plays with stereotypes of Indian women by painting them as animal-women. Not only does Sen's art work against stereotypes, but a critical commentary on capitalism and a countercapitalist ethos weave their way through her art. It is here that we encounter the method of creep. For instance, an artwork does not end with the frame in Sen's prolific exhibits.[3] Her images creep outside the frames, onto the walls. These creeping images make possession impossible, as the objects are no longer discrete. Consumers cannot ever be completely in possession of her art.

Sen provides us with a prism for un-technology or un–social media. Elaborating on her term *un–social media*, Sen says, "'Un-' allows me to playfully release myself (even if only momentarily) from the cycles of consumption and validation online, which are marked by the capitalist privatization of virtual space, surveillance technologies, e-commerce, and an overburdened economy of looking."[4] Her art, while using social media, also tricks it and consumers, by creating confusion over her online presence through multiple profiles and handles. Her art during the COVID lockdown in India and throughout the pandemic incessantly engaged with social media, such as her project *UnLOCKDOWN*. This project is a commentary and critique of the unceasing production expectations, even during a pandemic, a persistent scramble for "content generation."[5] Sen asks: "Are we now, as artists, simply supposed to generate content? And, in keeping with the markers of *good* content, are we to be relatable, viral, and clickable? To provide audiences with our critiques, capsuled in convenient word limits and shaped by an adequate aestheticization of politics?"[6] As her work delves into constructions of capitalism and the constraints that it places on the work of an artist to be continually marketable, Sen uses social media so that she can use it against itself.

Well before the lockdown, Sen had showcased technology and used it in her art. Consider *UnMYthU: UnKIND(s) Alternatives*, her 2018 multimedia performance with Alexa, Amazon's smart home assistant, for the Ninth Asia Pacific Triennial of Contemporary Art held in Brisbane, Australia.[7] This performance-exhibit was comprised of large drawings of animals and plants, and Sen conversed with Alexa alongside them, asking Alexa questions about the world at large. The conversation unraveled the contours of technology and the construction of artificial intelligence, alongside images of trees creeping through frames and animals feasting on other bodies. Creepy and macabre on multiple levels, the artwork also emphasizes an attempt to use and subvert technocreep by drawing out landscapes of human-animal-vegetal-technology in communion, and in tension. While this art predates COVID times, it acts as an ominous visualization of global contemporaneity devoured by capitalist technology that keeps profitable distinctions alive and makes invisible varied nuances of precarious living and death.

At the performance-exhibit, Sen appears brandishing a wand and asks Alexa, "How are you, Alexa?" This is the first of several questions that she asks Alexa, including whether Alexa has a mother tongue, whether she speaks Bengali and Spanish, and whether Alexa can love. These questions are asked in front of canvases portraying animals, boats, trees, and human bodies, merging and moving between frames. Sen asks Alexa the meaning of *refugee*, and Alexa provides the definition of *diffusion* as an answer. Translations, mistranslations, meanings, and mismeanings jostle together as Sen converses with technology in this artwork. In response to a question about waterboarding, Alexa provides a well-programmed answer that does not capture the nuances of pain or violence. Nor is Alexa able to answer a question about the boy from Syria who lost his life in the sea. Alongside asking questions in English, Sen also converses in gibberish that she articulates full of emotion and anguish. Sen's gibberish language does not contain the programmable and containable elements we encounter in Alexa's English. Throughout this performance-exhibit, we are left pondering the connections between language and technology, the language of technology, and its influence on ordering social reality. While technology reflects the language of society, and its power equations, it also mimetically projects the hierarchies of society as a programmed matter-of-fact status quo. Mainstream language reproduces the game of technology, and these continual reinforcements have dangerous consequences in making invisible the lives and languages that are not seen as profitable. Sen's questions for Alexa continue: "What is xenophobia?" "What is nationalism?" "What is an agreement?" "Alexa, do you love me?" While Alexa's programming helps her answer most of the questions, she is unable to answer the more personal or nuanced questions. Sen showcases the edges of technology or its thresholds, as well as the ways in which technology constantly reprograms itself to be marketable and relevant. Sen laughs vociferously, cries loudly, twirls her dress, and walks around in animated dissonance to the calm technology that answers questions about human tragedy in a matter-of-fact tone.

As we see in figure 11.1, Sen's buoyant performance uses skeleton images of humans, animals, and plants in entangled communion. The audience is privy to images of a tiny tiger, magnified bird, plants whose roots move in all directions, and the contours of human bodies. The anarchical scale-shifting questions the frames of the bodies, asking, Where do they begin and end? How is technology reading these scales? There are diagrams of all kinds superimposed on the images, as well as arrows, flow charts, and visual signifiers of classification, flow, and organization of categories. In this messy panorama, human, animal, and plant are shown as subjects for technology to organize and classify. And the mistranslations also emphasize its inability to do so.

The plant and animal images listen to the conversation between Sen and Alexa and testify to interconnections and disjointedness whereby technology cements categories of plant, animal, and human and can also blur the distinctions. As Elizabeth Povinelli succinctly writes in her book *Geontologies*, "Capital views all modes of existence as if they are vital *and* demands that not all modes of existence are the same from the point of view of extraction of value."[8] Povinelli's insights into the workings of capitalism draw out capitalism's adeptness to cement categories and selectively maneuver rights and recognition. In this context, Sen's questions to Alexa about whether Alexa is sexist or racist could be seen as an anguished cry within global contemporaneity to try holding technology responsible for the regimes of power that it always reflects and works within. Within the specificity of the questions is always superimposed a larger world view—the world that is neatly classified but a world that also

FIGURE II.I. *UnMYthU: UnKIND(s) Alternatives* by Mithu Sen

moves outside its projected frame. While plants and animals creep outside their frames in Sen's exhibit, technology—here in the form of Alexa and her world view—scrambles to define and contain them, stealthily creeping alongside.

While the performance was staged in 2018 in "pre"-pandemic times, Sen's art resonates through COVID times and challenges proclamations of a "post"-COVID world. Neither animals nor plants keep to their frames in Sen's art, and their border crossing illuminates the paradox of categories. The COVID virus itself defies categories. It is not, in scientific terms, alive. Yet it seems to have life: it evolves, adapts, and alters biological processes. If not a living thing, is it a form of technology? Like computer viruses, it creeps stealthily from body to body, from here to there, then to now, perforating the frames of plant, animal, life, and death and haunting them. Thinking through this performance, I ask: What, in fact, is technology? And how do we situate its functions in COVID times?

Exploring the Role of Technology in COVID Times

Thinking about technology unearths some interesting dilemmas. On a fundamental level, we are inspired to ask: What are the frontiers of technology and the human body? How do we mediate the relationship between the two, and where do animals and plants figure in this configuration? In paradigms of progress and capitalist modernization, technology is often seen as an artifact or tool. This way of thinking about technology as man-made art or craft has a long genealogy in Western thought. Aristotle, for instance, in the *Nicomachean Ethics* clearly distinguishes between *episteme* and *techne* in delineating forms of knowledge. While *episteme* referred to methods of knowledge, *techne* was skills and craft. For Aristotle, these are distinct in form, as spelled out in book 6 of the *Nicomachean Ethics*. Aristotelian presumptions that technology is the skilled use of artifact continue to dominate narratives of progress. However, more and more, cyborg realities, technocultures, and biotechnologies propel us toward thinking of technology as consistently entangled within the category of human, what Donna Haraway calls an "infolding." In *When Species Meet*, Haraway writes: "Technologies are not mediations, something in between us and another bit of the world. Rather, technologies are organs, full partners, in what [Maurice] Merleau-Ponty called 'infoldings of the flesh.' I like the word *infolding* better than *interface* to suggest the dance of world-making encounters. What happens in the folds is what is important. Infoldings of the flesh *are* worldly embodiment."[9]

Thinking about technologies as "organs" or as part of the "infolding of the flesh" anchors technology thinking within somatic structures and technological landscapes that undercut corporeal isolation of any kind. Opening the corporeal reach and embeddedness of technology must consider its race, gender, class, ability, and national contexts where the trade of organs or even the reading of flesh is dependent on power structures and political economy. For example, who gets to dance (and/or creep) in world-making, and under what conditions? As Sen's performance demonstrates, the language of power controls technology use that selectively includes and excludes entities in world-making. Ghosh's rendition of the invisibility of nonhuman communication also demonstrates the politics of "infolding of the flesh," where language as technology determines human arrogance. Thus, we are also led to ask, how are some humans used as technology in a manner that undermines their capacity for organic labor? While social stratifications influence the nature of technological encounters, another challenge when thinking about technology as an "infolding of the flesh" pertains to a visual simplification of folds as separate dimensions of matter that hold on to each other, inside out. In so many ways, our imagination of technologies, even when emphasizing their conjoined being-ness with humans, keeps imaging technology as distinct, whether inside or outside the human body. Anthropocentric border thinking dominates our corporeal landscape with a clear agenda to define the human body.

Thinking in broad strokes, our theories about technology as separate from corporeal bodies unfold into animal and plant framings. Humans are often credited with exclusive rights to language and technology. Such claims have over hundreds of years of political and social theorization. But even a cursory glance at actual animals and plants breaks such theorizations apart. The study of animal habitats and the construction of nests and burrows has long been used to demonstrate the technological adeptness of some animals, while a study of queer ecologies with an emphasis on the nonnormative sexual behavior of animals—behavior that belies reproduction as the sole rationale for sex—even shows animals' proclivity for using tools for sex.[10] Some animals, like an octopus or crow, are sensationalized and credited with vast tool-using prowess, of course with the added caveat that the understanding of *tool* has been largely anthropocentric. Plants are stationed below animals in stature; Jeffrey Nealon writes about the "abjection or exclusion of vegetable life within the voluminous work on nonhuman forms of life."[11] Plants have traversed an even longer route to be included in conversations about technology and language. While studies about plant sentience and communication patterns are now burgeoning in plant studies, the use of plants in technology has also intensified in recent years. From spinach that can detect explosives because of implanted nanotubes that display infrared light to plants that can remove soil contaminants, plants are active agents in bioengineering.[12] Plants are being used to generate electricity and are being planted with technology that enables faster growth patterns in the era of climate change.[13] Growing research on fungi, which are neither plant nor animal but instead serve as communicative mechanisms that demonstrate how plants use mycorrhizal networks. Thus, even though plants and animals use technology, and are used as technology, our reckoning of technology carries with it a curious anthropomorphic bias that always imagines technology as separate from the quintessentially human and places technology as helping to augment human prowess. Even though plants and animals use technology, they remain unseen unless we see some advantage of reckoning with them in our anthropocentric projects.

Technocreep as method means an attunement to technologies as motile, creeping processes that continually configure relations between man and machine, among other categories. In COVID times we see the proliferation of the virus everywhere, organic and inorganic, animal and machine, as seen in the spread of the coronavirus and the intensification of internet viruses as people spent more and more time online during the pandemic. Megan Squire talks about the similarities between the coronavirus and computer viruses by demonstrating how they both need a host to replicate and how both can be addressed by "disinfecting."[14] Much earlier in the temporal trajectory and before the full advent of the internet era, Jacques Derrida had written about viral technologies that could be both animal and machine. In *The Animal That Therefore I Am*, Derrida writes about the "animal-machine" that breaks apart the Cartesian dualism of animal/machine. He compares the animal-machine to a virus that can proliferate everywhere: "Neither animal nor nonanimal, neither organic nor inorganic, neither living nor dead, this potential invader is *like* a computer virus."[15] Working with continuums between animal-machine-technology, Derrida helps us imagine viral technologies that creep through the animal and vegetal.

COVID times are technology bound on multiple fronts and exemplify the methodology of technocreep as a stealthy creep through categories of social life. An explicit manifestation is the increased reliance on the internet for work and play, as the material infrastructures of social and economic life are relocated during the pandemic. As work shifted online, more companies invested in online services, and the internet saw an unprecedented volume of activity. A cyborgian reality has become more explicit in breaking boundaries of human and machine, alongside the humanization of the machine and the machination of the human. Zoom spaces have become familial and intimate, with occasional sights of pets and family members scuttling behind the screen and houseplants creeping from their planters. Human lives take on a dangerous mechanical pattern when separation between work and play and public and private are all subsumed under the same frame. Sen's play with scale, when presenting out-of-scale images, is omnipresent in our Zoom(ed) everyday realities, whether zooming in on Google Maps or being zoomed-out for the week. "Technostress" is becoming omnipresent as people are unable to regulate or curb the onslaught of technological intrusions in their lives.[16] The same technology that forms the material basis of relocated public and private infrastructures is also being used to curb and control the pandemic. Jobie Budd et al. showcase the mechanisms of "digital epidemiological surveillance" in the current global pandemic and show how digital technologies are being used in public health measures to counter the spread and contain the coronavirus.[17] They write, "Viruses know no borders and, increasingly, neither do digital technologies and data."[18]

When comprehending the role of technology in COVID times, we keep grappling with it as something coming from "outside" the human body but with an incessant proclivity to intrude and encroach into the psychosocial corporeal landscape. Would our analyses shift if we start thinking about the technological origins of the coronavirus and its technological "infolding of the flesh"? If the virus *is* technology, how do we think about the categories of human, animal, and plant within this technological matrix? There are various narratives about the origin of the coronavirus. While some confidently assert its zoonotic nature and spread through wet markets, others assert the lab accident theory.[19] The theory of the virus as bio-war also gained traction through the world.[20] While official investigations are ongoing to understand the genesis of the virus, the nature of a virus as neither animal, plant, nor human, and neither living nor dead, complicates our understanding of categories and the use of technology. Viruses need a host to proliferate, and they can live in plants, animals, humans, and bacteria alike. Considered good or bad based on their effect, viruses are of many kinds. They infold within other bodies, and in an ironic twist, all other bodies become their technology to proliferate. Indeed, COVID times make it impossible to ignore technocreep as methodology.

Technology in COVID times infolds with our animal thinking, too. To begin with, recall the exponential increase in animal sightings during the height of lockdown.[21] Animal videos went viral during this period. People reported seeing dolphins in Venice and Mumbai, pumas in Chile, and pink flamingos in France.[22] While some accounts were seen as hoaxes, globally people were surprised to see animals that seldom inhabit urban spaces openly. Social media, cell phones, cameras, and other forms of technology captured these images and circulated them through the world. Zoom meetings invariably include the presence of pets, and animal sightings of various kinds became a curious solace during the pandemic. While the presence of wild animals in urban spaces signaled a return of nature more than a spectacle, animals can also be a kind of technology against COVID. Horseshoe crabs are used in vaccine technology.[23] The animal's blue blood provides the lysate needed to detect contaminants and make a vaccine safe. The use of animals as technology to harness COVID, alongside the zoonotic origin narrative of COVID, makes animal lives and deaths central to the COVID landscape.

Newspapers in the early days of the pandemic reported widely about the decrease of pollution and carbon emissions. People traveled less, as many worked from home, and lockdown scenarios also ensured greater freedom for animals to rear their heads. This same phenomenon also accelerated poaching activities in wildlife refuges, endangering some animals more than others. Animals that rely on the human food chain, such as city rats and monkeys, were also affected in various ways.[24] Rats were reported to have become more aggressive in their search for food as restaurants closed during lockdown and there was less litter on the streets. While animals were variously impacted, houseplants have thrived in COVID times, as people bought more plants and spent more time with them during COVID.[25] Pet adoptions skyrocketed, and the sale of houseplants soared to an all-time high.[26] The natural world provided much solace as some people were forced to retreat inward and renew their relations to their immediate surroundings. Nature hikes became popular, rather than a visit to the movies, as people calculated safety variables. This back-to-nature phenomenon that accelerated in COVID times moves concomitantly with an intensification of technology use in social media, cell phones, and computers. How do we reconcile the two supposedly disparate frames: nature and technology? And how do we reconcile frighteningly different relations to nature and technology? I am thinking here about the crematoriums that ran out of wood in India in April 2021, in what Arundhati Roy describes as a "COVID catastrophe"; the forest department in Delhi invoked special permissions for the cutting of city trees.[27]

Back to nature gets spelled out very differently for people of different class, national, gender, and racial identities. Neda Atanasoski and Kalindi Vora, in their book *Surrogate Humanity*, write about "technoliberalism as the political alibi of present-day racial capitalism," explaining that "technological futures tied to capitalist development iterate a fantasy that as machines, algorithms, and artificial intelligence take over the dull, dirty, repetitive, and even reproductive labor performed by racialized, gendered, and colonized workers in the past, the full humanity of the (already) human subject will be freed for creative capacities."[28] Their book lucidly illuminates the uses of technology in COVID times. As we embrace nature with cell phone in hand so that we don't miss posting the perfect image of nature on social media, or when we marvel at animal sightings and like images that fuel their viral technological spread, and even when we think about spatial displacements and how the public and private are forever revoked for people economically displaced and homeless during the pandemic, Atanasoski and Vora add that racial capitalism undergirds each of these technological maneuvers. Race, class, gender, and nation mark questions of access, as well as the terms of interaction, between nature and technology. Returning to Sen's performance with Alexa, we see there, too, an exemplification of "technoliberalism" and its manipulations as they serve specific power regimes that profit from fixed meanings of objects and issues. While Alexa happily delivers predefined answers, it is not able to converse with tragedy, emotive nonlanguage, and creeping trees and bodies.

My intent in this chapter has been to outline the infoldings of COVID times that strengthen, subvert, use, and manipulate the categories of human, animal, plant, and machine. Attending to technocreep as method allows an attunement to the work of categories and to the subversions, infoldings, and perforations that are continuously taking place alongside one another. Drawn in broad strokes to signal frames and themes of co-relation, it remains important to decipher the constant use of categories in COVID times, while we also ceaselessly puncture them. Who is the human who cannot work or play without technology? What is technology that cannot play the human and undo "his" mastery? And what is the animal that proliferates through technology? How is the return to plants and nature realized in part as a question of technology? While these are undoubtedly broad questions in terms of the fields and conversations that it envelops, I remain convinced about the utility of thinking broadly so that we can decipher the full import of the workings of technology and who/ what is creeping through our lives in COVID times. Technocreep allows us to think about "the human" not as a natural category but as a creepy framing that works to enclose borders, of skin, breath, language, and technology. All along this framing remains an inevitable failure as logics of racial capitalism and colonialism meet spirited opponents in a creepy dance of infolded matter.

Thinking about Sen's exhibit from 2018 and our contemporary COVID times, I see in her work the contours of animals, plants, humans, and technology being conjoined and infolded to accentuate social and economic stratifications. Sen plays with the frames of animals, plants, humans, life, and death by questioning how technology mimics and controls categories, enlarges some, and obliterates others. While we celebrate animal sightings and forge new connections with plants and nature during COVID times, the growth of precarity and marginalization worsens for many, as people live and die unequally. Moving through this essay in broad strokes, purposefully broadening the frame to emphasize technocreep as methodology, I draw out the simultaneous accentuation and blurring of categories that characterize our contemporary moment. Hopefully this enables us to think about what and who is creeping through our world and in whose interest. While technocreep engulfs us all, some animals, plants, and humans creep outside their frames of enclosure.

The theme of creep centers anew our connected and entangled futures: animal, vegetal, and technology. I conclude this essay with the image of a creeping houseplant in figure 11.2 to explicitly illustrate how creepiness is part of the ontology of survival, alongside its presence as a mechanism of surveillance. The plant that slowly moves through its enclosure to spring forth with new leaves and tendrils and survey its surroundings to best anchor the growing tendrils illuminates the possibilities of creep as an ontological condition of flourishing. Much like Sen's trees, whose roots creep out of the frames of consumption, animals that creep toward safety, and technological subversions that sneak outside technocreep's intended objectives, thinking with creep as methodology draws our attention to diverse uses of strategies of survival and surveillance. Thinking

FIGURE 11.2. A creeping plant

broadly and through layers of intertwined materiality that help us visualize the paradoxical play of creep in forms around us, to snoop and violate boundaries, as well as stealthily cross out containments, we can think holistically about creeping nature-culture in COVID times and our entangled futures.

NOTES

Epigraph 1. Amitav Ghosh, "Brutes: Mediations on the Myth of the Voiceless," *Orion*, Autumn 2021, https://orionmagazine.org/article/brutes.

1. Chen, "Feminisms in the Air."

2. Chowdhury, "Precarity of Preexisting Conditions," 615.

3. See Chatterjee, "What Does It Mean to Be a Postcolonial Feminist?"

4. Zitzewitz and Sen, "'We Have Entered a Stage of Overproduction,'" 177.

5. Zitzewitz and Sen, "'We Have Entered a Stage of Overproduction,'" 181.

6. Zitzewitz and Sen, "'We Have Entered a Stage of Overproduction,'" 181–82.

7. Mithu Sen, *UnMYthU: UnKIND(s) Alternatives*, November 4, 2018, live performance at the Queensland Art Gallery/Gallery of Modern Art, Brisbane, commissioned for Ninth Asia Pacific Triennial of Contemporary Art, YouTube video, 25:16, www.youtube.com/watch?v=7oiL7AYWXNw.

8. Povinelli, *Geontologies*, 20.

9. Haraway, *When Species Meet*, 249.

10. See Alaimo, *Exposed*.

11. Nealon, *Plant Theory*, x.

12. Stuart Thompson, "Five Amazing Ways Plants Have Created New Technologies," *Conversation*, December 9, 2020, https://theconversation.com/five-amazing-ways-plants -have-created-new-technologies-69943.

13. Lucy Ingham, "Plant Power: The New Technology Turning Green Roofs into Living Power Plants," *Factor*, March 20, 2014, www.factor-tech.com/green-energy/1569 -plant-power-the-new-technology-turning-green-roofs-into-living-power-plants.

14. Owen Covington, "How COVID-19 Is Similar to the Viruses Trying to Infect Your Computer," Elon University, March 31, 2020, www.elon.edu/u/news/2020/03/31/how -covid-19-is-similar-to-the-viruses-trying-to-infect-your-computer.

15. Derrida, *Animal That Therefore I Am*, 39.

16. De, Pandey, and Pal, "Impact of Digital Surge during Covid-19 Pandemic," 2.

17. Budd et al., "Digital Technologies in the Public-Health Response to COVID-19."

18. Budd et al., "Digital Technologies in the Public-Health Response to COVID-19," 1189.

19. Haider et al., "COVID-19—Zoonosis or Emerging Infectious Disease?"

20. See Amy Maxmen, "US COVID Origins Report: Researchers Pleased with Scientific Approach," *Nature News*, August 27, 2021, www.nature.com/articles/d41586-021-02366-0.

21. See Chatterjee and Asher, "Animal Sightings and Citings under Covid Capitalism."

22. Radhika Chalasani, "Photos: Wildlife Roams during the Coronavirus Pandemic," ABC News, April 22, 2020, https://abcnews.go.com/International/photos-wildlife-roams -planets-human-population-isolates/story?id=70213431.

23. Carrie Arnold, "Horseshoe Crab Blood Is Key to Making a COVID-19 Vaccine—But the Ecosystem May Suffer," *National Geographic*, May 3, 2021, www .nationalgeographic.com/animals/article/covid-vaccine-needs-horseshoe-crab-blood.

24. Beatrice Jin, "How the Pandemic Has Changed the Natural World, Illustrated," *Politico*, May 19, 2020, www.politico.com/interactives/2020/coronavirus-pandemic -natural-world-illustrated.

25. Gideon Lasco, "How COVID-19 Is Changing People's Relationships with House-plants," *Sapiens*, September 17, 2020, www.sapiens.org/column/entanglements/covid-19 -houseplants.

26. Emily J. Sullivan, "Covid Lockdowns Turned Buying Plants into the Next Big Pandemic Trend—For Good Reason," NBC News, January 30, 2021, www.nbcnews.com /think/opinion/covid-lockdowns-turned-buying-plants-next-big-pandemic-trend-good -ncna1256223.

27. Arundhati Roy, "'We Are Witnessing a Crime against Humanity': Arundhati Roy on India's Covid Catastrophe," *Guardian*, April 28, 2021, www.theguardian.com/news /2021/apr/28/crime-against-humanity-arundhati-roy-india-covid-catastrophe.

28. Atanasoski and Vora, *Surrogate Humanity*, 4.

FIGURE E.I. *Close Your Eyes* by Sanaz Haghani.

NASSIM PARVIN

NEDA ATANASOSKI

Epilogue

Dreaming Feminist Futures

What can we see if we close our eyes and look for the evidence of things not seen? With eyes closed, dreamers see the evidence of things as light and shadow, as fading and emerging patterns. In a world privileging logic and the rational, what can dreams give us as evidence of what has been cast aside?

Working with Google's Deep Dream Generator—Alexander Mordintsev's computer vision program that utilizes convolutional neural networks to find images and enhance patterns—we created an artwork based on "creepy technology" as the sole prompt. The dream-like image it generated (fig. E.2) was surprising. There were no killer robots or hidden cameras. Instead, the image portrays the slow creep of vines and plants, foregrounded by lush grass. Tall, abandoned spiral towers shoot up from the landscape into a hazy pink and

FIGURE E.2. Google Deep Dream Generator image for the prompt "Creepy Technology," generated July 13, 2023

purple sky, but they blend into the naturescape. Plants dominate a world without humans or machines. Yet questions linger: Is this idyllic image actually a postapocalyptic vision? Or is it a utopian dream?

AI-generated images, like those of Deep Dream that are based on mining the vast archive of visual texts founds on the internet, have been compared to the art of the surrealism movement from the 1920s and 1930s. As Kathryn Johnson notes, this is both because of surrealism's focus on "dreams and desires" and because of its insistence that we "revise our notions of reality."[1] For example, dreams figure prominently in French writer and poet André Breton's "Manifesto of Surrealism." For him, dreams also play a significant role in unlocking what can't be perceived or thought when awake: "What I most enjoy contemplating about a dream is everything that sinks back below the surface in a waking state, everything I have forgotten about my activities in the course of the preceding day, dark foliage, stupid branches."[2] This description evokes the foliage and branches that creep over the constructed environment in the Deep Dream image of "creepy technology."

In an interview about the place of surrealism in the contemporary world, and as part of an exhibit on surrealism at the Design Museum in London, Deep Dream engineer Blaise Aguera y Arcas spoke to chief curator Justin McGuirk about AI and new forms of intelligence that can be used to produce dream-like imagery.[3] Aguera y Arcas notes that using language (prompting the AI) to create an image "allows us to cook up combinations that don't or can't exist in reality."[4] The "creation of impossible things" with AI is, for him, more about the collective unconscious than about an individual artist or creator.[5] This is because of the multiplicity of artistic styles represented in the ever-expanding archive of online images. At the same time, even as AI generates new connections and can help us see patterns and relationships that are otherwise unseen, we might observe the assumption in his statement is that art is otherwise about an individual artist or creator. This reflects Euro-American individualist notions of art in contrast to the view of artistic practice and inquiry as a mode of political engagement and action that is inherently social and situated.

Nonetheless, Aguera y Arcas shows how the processes of making AI-generated art move us away from an older, rationalist approach to "intelligence." Citing Eberhard Fetz, a neuroscientist at the University of Washington who compared Deep Dream's operations to what occurs to a human brain on acid trip, he observes that although Deep Dream wasn't trained on "psychedelic hippy art," it is no coincidence that the art has that quality. This is because "there is nothing rational about neural nets. . . . Rationality is a bit of a chimera. . . . [Gottfried Wilhelm] Leibnitz believed that we'd be able to rationalize everything. . . . That would be the triumph of rationality—to develop an algebra that could represent anything, and that could derive the rightness and wrongness of any position."[6] He explains that the dream of achieving perfect rationality was the momentum for the "old-fashioned AI movement." However, our own thoughts being imperfect and irrational, this dream could never come true, producing only "brittle systems."[7] In contrast, neural nets like Deep Dream are not "calculating a thing" or "following an algorithm" that rationally moves from A to B. Rather, they weigh and activate inputs in a way "not subject to logic in any sense that we can write down and understand."[8] They, in a sense, invent and create.

Rather than reinscribing a hierarchy of intelligence of an earlier moment of technoscience, in which teaching AI to play chess was considered a supreme achievement, Aguera y Arcas's articulation of intelligence is one that includes webs and relations between human creation, culture, and consciousness and evolving neural networks that can themselves dream the impossible. Still, other curators and critics are less enthusiastic about AI and nonhuman intelligence as creation. Rob Horning, for instance, writes that ChatGPT, Deep Dream,

DALL-E, and other open AI programs take surrealists' visions for a total turn away from human conscious agency to the extreme.

> When fed a textual prompt, GPT-3 . . . predicts what sentences should follow based on its statistical analysis of billions of words of text pulled from the internet. How it completes whatever prompt it's fed can be interpreted as a "social average" response, making it a kind of oblique search engine of the collective consciousness, liberated from any of the contextual social relations that would discipline what it produces. It doesn't experience inhibition or self-satisfaction. Thus it seems to fulfill Breton's wildest dreams for automatic writing, producing text that is estranged from human agency yet nonetheless has some perceivable sense to it that a reader can extract, or project on it.[9]

Put differently, AI cyclically reiterates normative values and privileges knowledges reflected in the vast online archive of images and text. Furthermore, it "'exponentially amplifies the knowledge shared by marketing experts with regard to our desires and fantasies' and is 'much quicker and much more efficient' in putting that knowledge to use."[10] Historian and curator Julian Stallabrass remarks that although "machine-generated works are nothing new" and since the "early 1970s [artists] wrote programmes that allowed computers to draw their own creations," what is new in the present moment of AI-generated art, "and undeniably impressive, is the scale and speed of this processing, the vast datasets on which it draws, and the hypnotic vision of an inhuman intelligence playing with human cultural techniques and material."[11] In doing so, emergent and expanding nonhuman intelligences amplify dominant modes of knowing and bolster actions that are aligned with the status quo and the most powerful.

The amplifications and generative connections that we observe with AI may be an occasion to recall IA, that is, intelligence amplified, a concept that, according to Alvin DMello, was introduced around the same time as AI: "In contrast to AI, which is a standalone system capable of processing information as well as or better than a human, IA is actually designed to complement and amplify human intelligence. IA has one big edge over AI: it builds on human intelligence that has evolved over millions of years, while AI attempts to build intelligence from scratch."[12] Turning our attention to what kinds of intelligences are culturally upheld puts into focus the practical and pragmatic issues that we face in the conception and integration of technologies. What kind of relations are we generating through our interaction with digital technologies, including the ones that are capable of generating text, images, and other media based on large databases? And, as Lucy Suchman evocatively asks, what are the human-machine configurations that we might strive for? How can we create alternate amplifications and juxtapositions that enable us to revise our notions of reality toward more just worlds and world-making?

Art, as we have shown throughout this book, has always been about making new connections and relationships and often in response to social, cultural, and political contexts—at times questioning and at other times reinforcing the dominant ethos or values. For example, artist Stephanie Dinkins worked with AI to produce images of Black women in a variety of affective states (for instance, smiling or crying), reflecting on the well-documented amplification of racial and gendered knowledge that represents the "social average" on AI platforms. In response to her word prompts, the "algorithm produced a pink-shaded humanoid shrouded by a black cloak."[13] Other artists, like Linda Dounia Rebaiz, have had similar experiences of racial and colonial distortions. Rebeiz "asked OpenAI's image generator, DALL-E 2, to imagine buildings in her hometown, Dakar. The algorithm produced arid desert landscapes and ruined buildings that Rebeiz said were nothing like the coastal homes in the Senegalese capital."[14] Many assessments of racism in technologies like facial recognition or AI place the blame on the history of mostly white men who were at the center of early machine learning research. More and more, studies speak about the "racial bias" that is baked into algorithms and neural networks, like the various image generators that are said to "automate" human racial bias. According to a 2023 *New York Times* article, "Major companies behind A.I. image generators—including OpenAI, Stability AI and Midjourney—have pledged to improve their tools. 'Bias is an important, industrywide problem,' Alex Beck, a spokeswoman for OpenAI, said ... adding that the company is continuously trying 'to improve performance, reduce bias and mitigate harmful outputs.'"[15]

Bias, of course, individualizes the problems of structural racism. It assumes that bias can be corrected to achieve a neutral norm—the latter a long-standing, if refuted, fantasy of science and technology. In this way, it is akin to discussions of generative AI hallucinations. Hallucinations are AI responses to queries that are not found in the training data. In other words, when AI confidently gives answers to queries that are not justified, such as providing a history of a place that doesn't exist, it is said to "hallucinate." Yet, as we and our contributors have argued throughout this book, the questions we ask about technology and AI should be not just about whether they are good or bad, hallucinating or factual, racist or "unbiased," but rather about the geopolitical, historical, and material contexts and relations in which they are produced and put to use. In fact, AI, as a mirror of our society, makes it easier than ever to see the "evidence of things not seen" or what most would rather not see—the legacies of settler colonial relations, structural racism, persistent misogyny and homophobia, and the investment in perpetuating racial capitalism, with Silicon Valley as one of its core engines. As Naomi Klein has written in an article that accuses Silicon Valley CEOs, rather than the generative AI, of hallucinating, "There is a world in which generative AI, as a powerful predictive research tool and a performer of tedious tasks, could indeed be marshaled to benefit humanity, other species and our shared home. But for that to happen, these technologies would need to be deployed inside a vastly different economic and social order than our own, one that had as its purpose the meeting of human needs and the protection of the planetary systems that support all life."[16]

In one sense, the language of bias and hallucination is useful because it challenges the assumptive authority of technology as objective and neutral. At the same time, the language of hallucination and bias is reductive because it assumes that technology is deviating from what is objective and neutral. In doing that, it reinforces the idea that nonbiased neutral information or knowledge is possible and desirable. A feminist approach toward the biases and hallucinations would recall the inherently situated nature of all knowledges.[17] We, however, invite our readers to move away from individualizing accusations of "bias" or "hallucination" that assume that a correction or expansion of data sets will sort the problem. Instead, we invite an orientation toward thinking of AI as shedding a spotlight on the colonial and racial foundations upon which technocaptalism is built. This, we posit, allows us to take AI as an occasion to reflect on existing relations and how to produce new ones. So, the main task is not that of removing bias and reinstating objectivity, but rather of embracing, thinking, and rethinking our inclinations. Could we see AI and technology as platforms for imaginative regeneration instead of wholesale objective truths?

Sanaz Haghani's artwork *Close Your Eyes* is generative of a new outlook on our embodied and always situated ways of knowing. What we see when we close our eyes is tightly connected to what is out there, yet also distant and detached from it. Asking what we still see when we close our eyes troubles our obsession with the visual and literal and refocuses our attention on the distortions and amplifications of light and shade. We, in effect, see our reality in a new light, just by closing our eyes. Taking that idea further, we might ask what we will see, and how differently we can see, if we engage all our senses. Trusting touch, intuition, and feminist and embodied intelligence is one of the lessons of technocreep. So is attentiveness to different time scales, ways of movement and change, and attunement ways of knowing that have historically been dismissed and disregarded—both our own and those of more-than-human beings. In this way, technocreep may indeed be an occasion for dreaming feminist futures.

NOTES

1. K. Johnson, "Absolute Reality," 7.

2. André Breton, "Manifestoes of Surrealism," 1924.

3. J. McGuirk, "Interview."

4. J. McGuirk, "Interview," 200.

5. J. McGuirk, "Interview," 200, 201.

6. J. McGuirk, "Interview," 202–3.

7. J. McGuirk, "Interview,"203.

8. J. McGuirk, "Interview," 204.

9. Rob Horning, "Word Processing," *Art in America*, April 18, 2022, www.artnews.com /art-in-america/features/word-processing-surrealism-artificial-intelligence-1234625624. Automatic writing and automatic art making are methods in which conscious control over the creation is suppressed to allow for the unconscious mind to express itself. Borrowing from the theories of Sigmund Freud, this was one of the methods embraced by the surrealists.

10. Horning, "Word Processing."

11. Stallabrass, "Sublime Calculation."

12. Alvin DMello, "Rise of the Humans," *Conversation*, October 12, 2015, https:// theconversation.com/rise-of-the-humans-intelligence-amplification-will-make-us-as -smart-as-the-machines-44767.

13. Zachary Small, "Black Artists Say A.I. Shows Bias, with Algorithms Erasing Their History," *New York Times*, July 4, 2023, www.nytimes.com/2023/07/04/arts/design /black-artists-bias-ai.html.

14. Small, "Black Artists."

15. Small, "Black Artists."

16. Naomi Klein, "AI Machines Aren't 'Hallucinating.' But Their Makers Are," *Guardian*, May 8, 2023, www.theguardian.com/commentisfree/2023/may/08/ai-machines -hallucinating-naomi-klein.

17. Haraway, "Situated Knowledges."

Bibliography

Agostinho, Daniela. "The Optical Unconscious of Big Data: Datafication of Vision and Care for Unknown Futures." *Big Data and Society* 6, no. 1 (2019): 1–10. https://doi.org/10.1177/2053951719826859.

Ahmed, Sara. *The Promise of Happiness.* Durham, NC: Duke University Press, 2010.

Alaimo, Stacy. *Exposed: Environmental Politics and Pleasures in Posthuman Times.* Minneapolis: University of Minnesota Press, 2016.

Alexander, Michelle. Foreword to *Prison by Any Other Name: The Harmful Consequences of Popular Reforms*, edited by Maya Schenwar and Victoria Law, ix–xvi. New York: New Press, 2020.

Allen, Tennille. "'I Didn't Let Everybody Come in My House': Exploring bell hooks' Notion of the Homeplace." *CLR James Journal* 17, no. 1 (2011): 75–101.

Amoore, Louise. *Cloud Ethics: Algorithms and the Attributes of Ourselves and Others.* Durham, NC: Duke University Press, 2020.

Amrute, Sareeta. "Of Techno-ethics and Techno-affects." *Feminist Review* 123, no. 1 (2019): 56–73. https://doi.org/10.1177/0141778919879744.

Andalibi, Nazanin. "Symbolic Annihilation through Design: Pregnancy Loss in Pregnancy-Related Mobile Apps." *New Media and Society* 23, no. 3 (2021): 613–31.

Andrejevic, Mark. "Automating Surveillance." *Surveillance and Society* 17, nos. 1–2 (2019): 7–13. https://doi:10.24908/ss.v17i1/2.12930.

Andrejevic, Mark. Foreword to Dubrofsky and Magnet, *Feminist Surveillance Studies*, ix–xix.

Antwi, Phanuel, Sarah Brophy, Helene Strauss, and Y-Dang Troeng. "Postcolonial Intimacies: Gatherings, Disruptions, Departures." *Interventions* 15, no. 1 (2013): 1–9. https://doi.org/10.1080/1369801X.2013.770994.

Anzaldúa, Gloria. *Borderlands/La Frontera: The New Mestiza.* San Francisco: Aunt Lute Books, 1987.

Arendt, Hannah. *The Human Condition.* Chicago: University of Chicago Press, 1998.

Atanasoski, Neda, and Kalindi Vora. *Surrogate Humanity: Race, Robots, and the Politics of Technological Futures.* Durham, NC: Duke University Press, 2019.

Baldwin, James. *Evidence of Things Not Seen*. New York: Holt, Rinehart and Winston, 1985.

Balsamo, Anne. *Designing Culture: The Technological Imagination at Work*. Durham, NC: Duke University Press, 2011.

Barassi, Veronica. "BabyVeillance? Expecting Parents, Online Surveillance and the Cultural Specificity of Pregnancy Apps." *Social Media and Society* 3, no. 2 (2017): 1–10. https://doi.org/10.1177/2056305117707188.

Barthes, Roland. *A Lover's Discourse: Fragments*. New York: Hill and Wang, 1996.

Bawaka Country, Sarah Wright, Sandie Suchet-Pearson, Kate Lloyd, Laklak Burarrwanga, Ritjilili Ganambarr, Merrkiyawuy Ganambarr-Stubbs, Banbapuy Ganambarr, Djawundil Maymuru, and Jill Sweeney. "Co-becoming Bawaka: Towards a Relational Understanding of Place/Space." *Progress in Human Geography* 40, no. 4 (2016): 455–75. https://doi.org:10.1177/0309132515589437.

Beaton, Brian. "Safety as Net Work: 'Apps against Abuse' and the Digital Labour of Sexual Assault Prevention." *Media Tropes* 5, no. 1 (2015): 105–24.

Beauchamp, Toby. *Going Stealth: Transgender Politics and US Surveillance Practices*. Durham, NC: Duke University Press, 2019.

Beittinger-Lee, Verena. *(Un)Civil Society and Political Change in Indonesia: A Contested Arena*. London: Routledge, 2009.

Belew, Kathleen. *Bring the War Home: The White Power Movement and Paramilitary America*. Cambridge, MA: Harvard University Press, 2018.

Beller, Jonathan. *The World Computer: Derivative Conditions of Racial Capitalism*. Durham, NC: Duke University Press, 2021.

Benjamin, Ruha, ed. *Captivating Technology: Race, Carceral Technoscience, and Liberatory Imagination in Everyday Life*. Durham, NC: Duke University Press, 2019.

Benjamin, Ruha. "Catching Our Breath: Critical Race STS and the Carceral Imagination." *Engaging Science, Technology, and Society*, no. 2 (2016): 145–56.

Benjamin, Ruha. *Race after Technology: Abolitionist Tools for the New Jim Code*. New York: Polity, 2019.

Benson, Robert W., and Raymond A. Biering. "Tenant Reports as an Invasion of Privacy: A Legislative Proposal." *Loyola of Los Angeles Law Review*, no. 12 (1978): 301–34.

Berlant, Lauren. *Cruel Optimism*. Durham, NC: Duke University Press, 2011.

Bhandar, Brenna. *Colonial Lives of Property: Law, Land, and Racial Regimes of Ownership*. Durham, NC: Duke University Press, 2018.

Bivens, Rena, and Amy Adele Hasinoff. "Rape: Is There an App for That? An Empirical Analysis of the Features of Anti-rape Apps." *Information, Communication and Society* 21, no. 8 (2018): 1050–67.

Bloch, Stefano. "Aversive Racism and Community-Instigated Policing: The Spatial Politics of Nextdoor." *Environment and Planning C: Politics and Space* 41, no. 1 (2022): 260–78. https://doi.org/10.1177/23996544211019754.

Block, Aaron, and Zach Aarons. *PropTech 101: Turning Chaos into Cash through Real Estate Innovation*. Charleston, SC: Advantage Media Group, 2019. www.proptech101.com.

Blomley, Nicholas. *Unsettling the City: Urban Land and the Politics of Property*. London: Routledge, 2004.

Bouk, Dan. *How Our Days Became Numbered*. Chicago: University of Chicago Press, 2015.

Boyd, S. B. *Challenging the Public/Private Divide.* Toronto: University of Toronto Press, 1997.

Brown, Marie V. B., and Albert L. Brown. Home security system utilizing television surveillance. US patent 3,482,037, filed August 1, 1966, and issued December 2, 1969. https://patents.google.com/patent/US3482037A/en.

Brown, Wendy. "Neo-liberalism and the End of Liberal Democracy." *Theory and Event* 7, no. 1 (2003). https://doi.org/10.1353/tae.2003.0020.

Browne, Simone. *Dark Matters: On the Surveillance of Blackness.* Durham, NC: Duke University Press, 2015.

Browne, Simone. "Digital Epidermalization: Race, Identity and Biometrics." *Critical Sociology* 36, no. 1 (2010): 131–50. https://doi.org/10.1177/0896920509347144.

Bruce, La Marr Jurelle. *How to Go Mad without Losing Your Mind: Madness and Black Radical Creativity.* Durham, NC: Duke University Press, 2021.

Bucher, Taina. *If . . . Then: Algorithmic Power and Politics.* Oxford: Oxford University Press, 2018.

Budd, Jobie, Benjamin S. Miller, Erin M. Manning, Vasileios Lampos, Mengdie Zhuang, Michael Edelstein, Geraint Rees, et al. "Digital Technologies in the Public-Health Response to COVID-19." *Nature Medicine*, no. 26 (2020): 1183–92. https://doi.org/10.1038/s41591-020-1011-4.

Bui, Long T. *Model Machines: A History of the Asian as Automaton.* Philadelphia: Temple University Press, 2022.

Buolamwini, Joy, and Timnit Gebru. "Gender Shades: Intersectional Accuracy Disparities in Commercial Gender Classification." *Proceedings of Machine Learning Research*, no. 81 (2018): 1–15. Conference on Fairness, Accountability, and Transparency.

Buttigieg, Joseph A. "Gramsci on Civil Society." *boundary 2* 22, no. 3 (1995): 1–32.

Byrd, Jodi, Chandan Reddy, Alyosha Goldstein, and Jodi Melamed. "Predatory Value: Economies of Dispossession and Disturbed Relationalities." *Social Text* 36, no. 2(135) (2018): 1–18.

Campolo, Alexander, and Kate Crawford. "Enchanted Determinism: Power without Responsibility in Artificial Intelligence." *Engaging Science, Technology, and Society*, no. 6 (2020): 1–19. https://doi.org/10.17351/ests2020.277.

Campt, Tina M. *Listening to Images.* Durham, NC: Duke University Press, 2017.

Capan, Faruk. "Why Amazon Device Is a Gift for Healthcare." *Medical Marketing and Media* 51, no. 1 (2016): 20.

Carrara, Sandro. "Body Dust: Well Beyond Wearable and Implantable Sensors." *IEEE Sensors Journal* 21, no. 11 (2021): 12398–406. https://doi.org/10.1109/JSEN.2020.3029432.

Ceccaroni, Luigi, Anne Bowser, and Peter Brenton. "Civic Education and Citizen Science: Definitions, Categories, Knowledge Representation." In *Analyzing the Role of Citizen Science in Modern Research*, edited by Luigi Ceccaroni and Jaume Piera, 1–23. Hershey, PA: IGI Global, 2017.

Chaar López, Iván. "Alien Data: Immigration and Regimes of Connectivity in the United States." *Critical Ethnic Studies* 6, no. 2 (2021). https://doi.org/10.5749/CES.0602.lopez.

Chaar López, Iván. *The Cybernetic Border.* Durham, NC: Duke University Press, 2024.

Chaar López, Iván. "Sensing Intruders: Race and the Automation of Border Control."
American Quarterly 71, no. 2 (2019): 495–518.

Chakravartty, Paula, and Denise Ferreira da Silva. "Accumulation, Dispossession, and Debt:
The Racial Logic of Global Capitalism." *American Quarterly* 64, no. 3 (2012): 361–85.

Chambers, Simone, and Jeffrey Kopstein. "Bad Civil Society." *Political Theory* 29, no. 6
(2001): 837–65.

Chatterjee, Sushmita. "What Does It Mean to Be a Postcolonial Feminist? The Artwork
of Mithu Sen." *Hypatia* 31, no. 1 (2016): 22–40.

Chatterjee, Sushmita, and Kiran Asher. "Animal Sightings and Citings under Covid Capitalism: Beyond Liberal Sentimentalism." *Feminist Studies* 47, no. 3 (2021): 599–626.

Chávez, Alex E. "Gender, Ethno-nationalism, and the Anti-Mexicanist Trope." *Journal of
American Folklore* 134, no. 531 (Winter 2021): 3–24.

Chavez, Leo. *The Latino Threat: Constructing Immigrants, Citizens, and the Nation.*
2nd ed. Stanford, CA: Stanford University Press, 2013.

Chen, Mel Y. "Feminisms in the Air." *Signs* 47, no. 1 (2021): 22–29. www.journals
.uchicago.edu/doi/abs/10.1086/715733.

Chilvers, Jason, and Matthew Kearnes. "Remaking Participation in Science and Democracy."
Science, Technology, and Human Values 45, no. 3 (2020): 347–80.

Chmieliauskas, Sigitas, Eimantas Mundinas, Dmitrij Fomin, Gerda Andriuskeviciute, Sigitas Laima, Eleonora Jurolaic, Jurgita Stasiuniene, and Algimantas Jasulaitis. "Sudden
Deaths from Positional Asphyxia: A Case Report." *Medicine* 97, no. 24 (2018): 1–5.

Chowdhury, Elora Halim. "The Precarity of Preexisting Conditions." *Feminist Studies*
46, no. 3 (2020): 615–25.

Christin, Angèle. "The Ethnographer and the Algorithm: Beyond the Black Box." *Theory
and Society*, no. 49 (2020): 897–918. https://doi.org/10.1007/s11186-020-09411-3.

Chun, Wendy H. K. *Discriminating Data: Correlation, Neighborhoods, and the New
Politics of Recognition.* Cambridge, MA: MIT Press, 2021.

Chun, Wendy H. K. *Updating to Remain the Same: Habitual New Media.* Cambridge,
MA: MIT Press, 2017.

Collins, Patricia Hill. *Black Feminist Thought: Knowledge, Consciousness, and the Politics
of Empowerment.* New York: Routledge, 2002.

Cooper, Caren, and Bruce Lewenstein. "Two Meanings of Citizen Science." In *The Right-
ful Place of Science: Citizen Science,* edited by Darlene Cavalier and Eric B. Kennedy,
51–62. Tempe, AZ: Consortium for Science, Policy and Outcomes, 2016.

Cottom, Tressie McMillan. *Thick: And Other Essays.* New York: New Press, 2018.

Cowan, Ruth Schwartz. *More Work for Mother: The Ironies of Household Technology from
the Open Hearth to the Microwave.* New York: Basic Books, 1983.

Crary, Jonathan. *24/7: Late Capitalism and the Ends of Sleep.* London: Verso, 2014.

Crawford, Kate, Jessa Lingel, and Tero Karppi. "Our Metrics, Ourselves: A Hundred
Years of Self-Tracking from the Weight Scale to the Wrist Wearable Device." *European
Journal of Cultural Studies* 18, nos. 4–5 (2015): 479–96.

Cummins, Nicholas, Stefan Scherer, Jarek Krajewski, Sebastian Schnieder, Julien Epps,
and Thomas F. Quatieri. "A Review of Depression and Suicide Risk Assessment Using
Speech Analysis." *Speech Communication*, no. 71 (2015): 10–49. https://doi.org/10.1016
/j.specom.2015.03.004.

Davis, Dana-Ain. *Reproductive Injustice: Racism, Pregnancy, and Premature Birth*. New York: NYU Press, 2019.

Davis, Jenny. *How Artifacts Afford: The Power and Politics of Everyday Things*. Cambridge, MA: MIT Press, 2020.

De, Rahul, Neena Pandey, and Abhipsa Pal. "Impact of Digital Surge during Covid-19 Pandemic: A Viewpoint on Research and Practice." *International Journal of Information Management*, no. 55 (2020): 102171. https://doi.org/10.1016/j.ijinfomgt.2020 .102171.

Deer, Sarah. *The Beginning and End of Rape: Confronting Sexual Violence in Native America*. Minneapolis: University of Minnesota Press, 2015.

De León, Jason. *The Land of Open Graves: Living and Dying on the Migrant Trail*. Berkeley: University of California Press, 2015.

Delfanti, Alessandro, and Bronwyn Frey. "Humanly Extended Automation or the Future of Work Seen through Amazon Patents." *Science, Technology, and Human Values* 46, no. 3 (2021): 655–82. https://doi.org/10.1177/0162243920943665.

Dellinger, Jolynn, and Stephanie K. Pell. "Bodies of Evidence: The Criminalization of Abortion and Surveillance of Women in a Post-Dobbs World." *Duke Journal of Constitutional Law and Public Policy*, 19, no. 1 (2024). Duke Law School Public Law and Legal Theory Series no. 2023-66. https://scholarship.law.duke.edu/djclpp/vol19/iss1/1/.

Derrida, Jacques. *The Animal That Therefore I Am*. Edited by Marie-Louise Mallet. Translated by David Wills. New York: Fordham University Press, 2008.

Díaz-Barriga, Miguel, and Margaret E. Dorsey. *Fencing In Democracy: Border Walls, Necrocitizenship, and the Security State*. Durham, NC: Duke University Press, 2020.

D'Ignazio, Catherine, and Lauren F. Klein. *Data Feminism*. Cambridge, MA: MIT Press, 2020.

Dillon, Sarah. "The Eliza Effect and Its Dangers: From Demystification to Gender Critique." *Journal for Cultural Research* 24, no. 1 (2020): 1–15.

"Do I Sound Sick?" *Lancet Digital Health* 3, no. 9 (2021): e534. https://doi.org/10.1016 /S2589-7500(21)00182-5.

Dubrofsky, Rachel E., and Shoshana Amielle Magnet, eds. *Feminist Surveillance Studies*. Durham, NC: Duke University Press, 2015.

Eidsheim, Nina Sun. *Sensing Sound: Singing and Listening as Vibrational Practice*. Durham, NC: Duke University Press, 2015.

Eisenhut, Katharina, Ela Sauerborn, Claudia García-Moreno, and Verina Wild. "Mobile Applications Addressing Violence against Women: A Systematic Review." *BMJ Global Health* 5, no. 4 (2020): e001954.

Elwood, Sarah. "Digital Geographies, Feminist Relationality, Black and Queer Code Studies: Thriving Otherwise." *Progress in Human Geography* 45, no. 2 (2021): 209–28.

Elyachar, Julia. "Phatic Labor, Infrastructure, and the Question of Empowerment in Cairo." *American Ethnologist* 37, no. 3 (2010): 452–64. https://doi.org/10.1111/j.1548 -1425.2010.01265.x.

Epstein, Steven. *Impure Science: AIDS, Activism, and the Politics of Knowledge*. Berkeley: University of California Press, 1996.

Erevelles, Nirmala. "'Coming Out Crip' in Inclusive Education." *Teachers College Record* 113, no. 10 (2011): 2155–85.

Erevelles, Nirmala. *Disability and Difference in Global Contexts: Enabling a Transformative Body Politic*. New York: Springer, 2011.

Eubanks, Virginia. *Automating Inequality: How High-Tech Tools Profile, Police, and Punish the Poor*. New York: St. Martin's, 2018.

Eubanks, Virginia. *Digital Dead End: Fighting for Social Justice in the Information Age*. Cambridge, MA: MIT Press, 2011.

Falcon, Sylvanna. "Rape as a Weapon of War: Advancing Human Rights for Women at the US-Mexico Border." *Social Justice* 28, no. 2 (2001): 31–50.

Fedorova, Ksenia. *Tactics of Interfacing: Encoding Affect in Art and Technology*. Cambridge, MA: MIT Press, 2020.

Ferrer, Alexander. *Beyond Wall Street Landlords: How Private Equity in the Rental Market Makes Housing Unaffordable, Unstable, and Unhealthy*. Los Angeles: Strategic Action for a Just Economy, 2021. www.saje.net/wp-content/uploads/2021/03/Final_A-Just-Recovery-Series_Beyond_Wall_Street.pdf.

Fields, Desiree. "Automated Landlord: Digital Technologies and Post-crisis Financial Accumulation." *Environment and Planning A: Economy and Space* 54, no. 1 (2022): 160–81.

Fisher, Mark. *The Weird and the Eerie*. London: Watkins Media, 2017.

Fox, Sarah, and Franchesca Spektor. "Hormonal Advantage: Retracing Exploitative Histories of Workplace Menstrual Tracking." *Catalyst: Feminism, Theory, Technoscience* 7, no. 1 (2021): 1–23. https://catalystjournal.org/index.php/catalyst/article/view/34506.

Fraser, Nancy. "Sex, Lies, and the Public Sphere: Some Reflections on the Confirmation of Clarence Thomas." *Critical Inquiry* 18, no. 3 (1992): 595–612.

Fraser, Nancy, and Kate Nash. *Transnationalizing the Public Sphere*. Cambridge, MA: Polity, 2014.

Fritsch, Kelly. "Cripping Neoliberal Futurity: Marking the Elsewhere and Elsewhen of Desiring Otherwise." *Feral Feminisms*, no. 5 (2016): 11–26.

Gabrys, Jennifer. "Programming Nature as Infrastructure in the Smart Forest City." *Journal of Urban Technology* 29, no. 1 (2022): 13–19.

Gabrys, Jennifer. "Smart Forests and Data Practices: From the Internet of Trees to Planetary Governance." *Big Data and Society* 7, no. 1 (2020): 1–10. https://doi.org/10.1177/2053951720904871.

Gibbons, Joseph. "'Placing' the Relation of Social Media Participation to Neighborhood Community Connection." *Journal of Urban Affairs* 42, no. 8 (2020): 1262–77.

Gieseking, Jack. "Operating Anew: Queering GIS with Good Enough Software." *Canadian Geographer/Le Géographe Canadien* 62, no. 1 (2018): 55–66.

Gillespie, Tarleton. "The Relevance of Algorithms." In *Media Technologies: Essays on Communication, Materiality, and Society*, edited by Tarleton Gillespie, Pablo J. Boczkowski, and Kirsten A. Foot, 167–93. Cambridge, MA: MIT Press, 2014.

Gilmore, Ruth Wilson. *Golden Gulag: Prisons, Surplus, Crisis, and Opposition in Globalizing California*. Berkeley: University of California Press, 2007.

Gitelman, Lisa, ed. *"Raw Data" Is an Oxymoron*. Cambridge, MA: MIT Press, 2013.

Glenn, Evelyn Nakano. "From Servitude to Service Work: Historical Continuities in the Racial Division of Paid Reproductive Labor." *Signs* 18, no. 1 (1992): 1–43. https://doi.org/10.1086/494777.

Glück, Zoltán, and Setha Low. "A Sociospatial Framework." *Anthropological Theory* 17, no. 3 (2017): 281–96.

Goodman, Nanette, Michael Morris, Kelvin Boston, and Donna Walton. *Financial Inequality: Disability, Race and Poverty in America*. Washington, DC: National Disability Institute, 2017.

Goudzwaard, Bob, and Harry de Lange. *Beyond Poverty and Affluence: Toward an Economy of Care with a Twelve-Step Program for Economic Recovery*. Geneva: WCC, 1992.

Gould, Corrina. "Ohlone Geographies." In *Counterpoints: A San Francisco Bay Area Atlas of Displacement and Resistance*, edited by Anti-Eviction Mapping Project, 71–75. Oakland, CA: PM Press, 2021.

Gregg, Melissa. *Counterproductive: Time Management in the Knowledge Economy*. Durham, NC: Duke University Press, 2018.

Grewal, Inderpal. *Saving the Security State: Exceptional Citizens in Twenty-First-Century America*. Durham, NC: Duke University Press, 2017.

Gumbs, Alexis Pauline. "Undrowned: Black Feminist Lessons from Marine Mammals." *Soundings*, no. 78 (2021): 20–37. https://doi.org/10.3898/soun.78.01.2021.

Haider, Najmul, Peregrine Rothman-Ostrow, Abdinasir Yusuf Osman, Lia Barbara Arruda, Laura Macfarlane-Berry, Linzy Elton, Margaret J. Thomason, et al. "COVID-19— Zoonosis or Emerging Infectious Disease?" *Frontiers in Public Health,* no. 8 (2020): 1–8. https://doi.org/10.3389/fpubh.2020.596944.

Hall, Stuart. *Policing the Crisis: Mugging, the State, and Law and Order*. New York: Macmillan, 1978.

Hamraie, Aimi. *Building Access: Universal Design and the Politics of Disability*. Minneapolis: University of Minnesota Press, 2017.

Hamraie, Aimi. "Universal Design and the Problem of 'Post-disability' Ideology." *Design and Culture* 8, no. 3 (2016): 285–309. https://doi.org/10.1080/17547075.2016.1218714.

Hamraie, Aimi, and Kelly Fritsch. "Crip Technoscience Manifesto." *Catalyst: Feminism, Theory, Technoscience* 5, no. 1 (2019): 1–33. https://doi.org/10.28968/cftt.v5i1.29607.

Haraway, Donna. *The Companion Species Manifesto: Dogs, People, and Significant Otherness*. Vol. 1. Chicago: Prickly Paradigm, 2003.

Haraway, Donna. *Modest_Witness@Second_Millennium.FemaleMan©_Meets _Oncomouse™*. New York: Routledge, 1997.

Haraway, Donna. "Situated Knowledges: The Science Question in Feminism and the Privilege of Partial Perspective." *Feminist Studies* 14, no. 3 (1988): 575–99. https://doi .org/10.2307/3178066.

Haraway, Donna. *When Species Meet*. Minneapolis: University of Minnesota Press, 2008.

Harding, Sandra. *Objectivity and Diversity: Another Logic of Scientific Research*. Chicago: University of Chicago Press, 2015.

Harris, Cheryl. "Whiteness as Property." *Harvard Law Review* 106, no. 8 (1993): 1710–91.

Harris, Nikita Mae, Robert Lindeman, Clara Shui Fern Bah, Daniel Gerhard, and Simon Hoermann. "Eliciting Real Cravings with Virtual Food: Using Immersive Technologies to Explore the Effects of Food Stimuli in Virtual Reality." *Frontiers in Psychology* 14, no. 956585 (April 2023). www.ncbi.nlm.nih.gov/pmc/articles/PMC10149689.

Helmreich, Stefan. "An Anthropologist Underwater: Immersive Soundscapes, Submarine Cyborgs, and Transductive Ethnography." *American Ethnologist* 34, no. 4 (2007): 621–41. https://doi.org/10.1525/ae.2007.34.4.621.

Helmreich, Stefan. "Transduction." In *Keywords in Sound Studies*, edited by David Novak and Matt Sakakeeny, 222–31. Durham, NC: Duke University Press, 2015.

Henne, Kathryn, Renee Shelby, and Jenna Harb. "The Datafication of #MeToo: Whiteness, Racial Capitalism, and Anti-violence Technologies." *Big Data and Society* 8, no. 2 (2021): 1–14.

Henriques, Julian. *Sonic Bodies: Reggae Sound Systems, Performance Techniques, and Ways of Knowing*. London: Bloomsbury, 2011.

Herring, Chris. "Complaint-Oriented Policing: Regulating Homelessness in Public Space." *American Sociological Review* 84, no. 5 (2019): 769–800.

Hester, Helen. *Xenofeminism*. Cambridge, UK: Polity, 2018.

Hester, Helen, and Nick Srnicek. "The Crisis of Social Reproduction and the End of Work." In BBVA, *The Age of Perplexity: Rethinking the World We Knew*, 335–51. New York: Penguin Random House, 2018. www.bbvaopenmind.com/en/articles/the-crisis -of-social-reproduction-and-the-end-of-work.

Hirshbein, Laura Davidow. "Masculinity, Work, and the Fountain of Youth: Irving Fisher and the Life Extension Institute, 1914–31." *Canadian Bulletin of Medical History* 16, no. 1 (1999): 89–124.

Hobart, Hiʻilei Julia Kawehipuaakahaopulani, and Tamara Kneese. "Radical Care: Survival Strategies for Uncertain Times." *Social Text* 38, no. 1 (142) (2020): 1–16. https:// doi.org/10.1215/01642472-7971067.

hooks, bell. "Homeplace: A Site of Resistance." In *Yearning: Race, Gender, and Cultural Politics*, 41–50. Cambridge, UK: South End, 1990.

Hsieh, Jennifer C. "Piano Transductions: Music, Sound and Noise in Urban Taiwan." *Sound Studies* 5, no. 1 (2019): 4–21. https://doi.org/10.1080/20551940.2018 .1564459.

Hua, Julietta, and Kasturi Ray. *Spent behind the Wheel: Drivers' Labor in the Uber Economy*. Minneapolis: University of Minnesota Press, 2021.

Hughson, Jo-Anne Patricia, J. Oliver Daly, Robyn Woodware-Kron, John Hajek, and David Story. "The Rise of Pregnancy Apps and the Implications for Culturally and Linguistically Diverse Women." *JMIR mHealth and uHealth* 6, no. 11 (2018): 1–19. https://mhealth.jmir.org/2018/11/e189.

Hull, Gordon, and Frank Pasquale. "Toward a Critical Theory of Corporate Wellness." *BioSocieties* 13, no. 1 (2018): 190–212.

Humphreys, Lee. *The Qualified Self*. Cambridge, MA: MIT Press, 2018.

Inoue, Miyako. "Word for Word: Verbatim as Political Technologies." *Annual Review of Anthropology* 47, no. 1 (2018): 217–32. https://doi.org/10.1146/annurev-anthro-102116 -041654.

Iralu, Elspeth. "Putting Indian Country on the Map: Indigenous Practices of Spatial Justice." *Antipode* 53, no. 5 (2021): 1485–502.

Irani, Lilly. "The Cultural Work of Microwork." *New Media and Society* 17, no. 5 (2015): 720–39.

Irani, Lilly. "Difference and Dependence among Digital Workers: The Case of Amazon Mechanical Turk." *South Atlantic Quarterly* 114, no. 1 (2015): 225–34. https://doi.org /10.1215/00382876-2831665.

Irwin, Alan. *Citizen Science: A Study of People, Expertise and Sustainable Development.* London: Routledge, 1995.

Jacobs, Jane. *The Death and Life of Great American Cities.* New York: Vintage Books, 1961.

Jacques-Tiura, Angela J., Rifky Tkatch, Antonia Abbey, and Rhiana Wegner. "Disclosure of Sexual Assault: Characteristics and Implications for Posttraumatic Stress Symptoms among African American and Caucasian Survivors." *Journal of Trauma and Dissociation* 11, no. 2 (2010): 186.

JafariNaimi, Nassim, Lisa Nathan, and Ian Hargraves. "Values as Hypotheses: Design, Inquiry, and the Service of Values." *Design Issues* 31, no. 4 (2015): 91–104.

Jakobson, Roman. "Closing Statements: Linguistics and Poetics." In *Styles in Language,* edited by Thomas A. Sebeok, 350–77. Cambridge, MA: MIT Press, 1960.

Jasanoff, Sheila. *Design on Nature: Science and Democracy in Europe and the United States.* Princeton, NJ: Princeton University Press, 2005.

Jefferson, Brian. *Digitize and Punish: Racial Criminalization in the Digital Age.* Minneapolis: University of Minnesota Press, 2020.

Jen, Gish. *The Resisters.* New York: Knopf Doubleday, 2020.

Jimenez, Maria. *Humanitarian Crisis: Migrant Deaths at the US-Mexico Border.* New York: American Civil Liberties Union of San Diego and Imperial Counties and Mexico's National Commission of Human Rights, 2009.

Johnson, Benjamin. *Revolution in Texas: How a Forgotten Rebellion and Its Bloody Suppression Turned Mexicans into Americans.* New Haven, CT: Yale University Press, 2005.

Johnson, Carolyn Y. "Long Overlooked by Science, Pregnancy Is Finally Getting the Attention It Deserves." *Washington Post,* March 3, 2019. https://www.washingtonpost .com/national/health-science/long-overlooked-by-science-pregnancy-is-finally-getting -attention-it-deserves/2019/03/06/a29ae9bc-3556-11e9-af5b-b51b7ff322e9_story.html.

Johnson, Kathryn. "Absolute Reality." In K. Johnson, *Surrealism and Design Now,* 7–11.

Johnson, Kathryn, ed. *Surrealism and Design Now: From Dalí to AI.* London: Design Museum Publishing, 2022.

Kafer, Alison. *Feminist, Queer, Crip.* Bloomington: University of Indiana Press, 2013.

Kak, Amba, and Sarah Myers West. *AI Now 2023 Landscape: Confronting Tech Power.* New York: AI Now Institute, 2023. https://ainowinstitute.org/2023-landscape.

Kaplan, Amy. *The Anarchy of Empire in the Making of US Culture.* Cambridge, MA: Harvard University Press, 2002.

Kasmir, Seton Paul, and Joseph Frank Scalisi. Doorbell Communication Systems and Methods. U.S. Patent no. 9,179,109 B1, filed May 30, 2015, and issued November 3, 2015. https://patents.google.com/patent/US9055202B1/en.

Keenan, Thomas. *Technocreep: The Surrender of Privacy and the Capitalization of Intimacy.* Vancouver, BC: Greystone Books, 2014.

Kelleher, John D. *Deep Learning.* Cambridge, MA: MIT Press, 2019.

Keller, Evelyn Fox. *A Feeling for the Organism: The Life and Work of Barbara McClintock.* London: Macmillan, 1984.

Kimura, Aya Hirata. *Radiation Brain Moms and Citizen Scientists: The Gender Politics of Food Contamination after Fukushima*. Durham, NC: Duke University Press, 2016.

Kneese, Tamara. "A Responsible Death: Life Insurance from Mortality Tables to Wearables." In *The New Death*, edited by Shannon Lee Dawdy and Tamara Kneese, 73–104. Santa Fe, NM: School for Advanced Research Press, 2022.

Koslowski, Rey, and Marcus Schulzke. "Drones along Borders: Border Security UAVs in the United States and the European Union." *International Studies Perspectives* 19, no. 4 (2018): 305–24.

Kurwa, Rahim. "Building the Digitally Gated Community: The Case of Nextdoor." *Surveillance and Society* 17, nos. 1–2 (2019): 111–17. https://doi.org/10.24908/ss.v17i1/2.12927.

Lambright, Katie. "Digital Redlining: The Nextdoor App and the Neighborhood of Make-Believe." *Cultural Critique* 103, no. 1 (2019): 84–90. https://doi.org/10.1353/cul.2019.0023.

Langer, Markus, and Cornelius J. König. "Introducing and Testing the Creepiness of Situation Scale (CRoSS)." *Frontiers in Psychology*, no. 9 (2018): 1–17. www.frontiersin.org/articles/10.3389/fpsyg.2018.02220/full.

Le Corbusier. *Towards a New Architecture*. New York: Dover, 1986. First published 1931 by John Rodker (London).

Levario, Miguel A. "Home Guard: State-Sponsored Vigilantism and Violence in the Texas-Mexico Borderlands." In *Border Policing: A History of Enforcement and Evasion in North America*, edited by Holly M. Karibo and George T. Díaz, 129–43. Austin: University of Texas Press, 2020.

Levy, Karen. "Intimate Surveillance." *Idaho Law Review* 51, no. 3 (2015): 679–93. https://digitalcommons.law.uidaho.edu/idaho-law-review/vol51/iss3/5.

Lewis, Jason Edward, Noelani Arista, Archer Pechawis, and Suzanne Kite. "Making Kin with the Machines." *Journal of Design and Science*, no. 3.5 (2018). https://doi.org/10.21428/bfafd97b.

Liboiron, Max. *Pollution Is Colonialism*. Durham, NC: Duke University Press, 2021.

Lingel, Jessa, and Kate Crawford. "'Alexa, Tell Me about Your Mother': The History of the Secretary and the End of Secrecy." *Catalyst: Feminism, Theory, Technoscience* 6, no. 1 (2020): 1–23.

Lipsitz, George. "The Possessive Investment in Whiteness: Racialized Social Democracy and the 'White' Problem in American Studies." *American Quarterly* 47, no. 3 (1995): 369–87.

Liu, Chuncheng. "Seeing like a State, Enacting like an Algorithm: (Re)Assembling Contact Tracing and Risk Assessment during the COVID-19 Pandemic." *Science, Technology, and Human Values* 47, no. 4 (2022): 698–725. https://doi.org/10.1177/01622439211021916.

Logan, Wayne A. "Crowdsourcing Crime Control." *Texas Law Review* 99, no. 1 (2020): 137–63.

Lorde, Audre. *Uses of the Erotic: The Erotic as Power*. Brooklyn, NY: Out and Out Books, 1978. www.centralcontemporaryarts.org/audre-lorde-the-erotic-as-power.

Loukissas, Yanni Alexander. *All Data Are Local: Thinking Critically in a Data-Driven Society*. Cambridge, MA: MIT Press, 2019.

Low, Setha M. "Fortification of Residential Neighbourhoods and the New Emotions of Home." *Housing, Theory and Society* 25, no. 1 (2008): 47–65. https://doi.org/10.1080/14036090601151038.

Lupton, Deborah. "Quantified Sex: A Critical Analysis of Sexual and Reproductive Self-Tracking Using Apps." *Culture, Health, and Sexuality* 17, no. 4 (2015): 440–53.

Lyon, David. *Surveillance Society: Monitoring Everyday Life*. Buckingham: Open University Press, 2011.

MacPherson, C. B. *The Political Theory of Possessive Individualism: Hobbes to Locke.* Oxford: Oxford University Press, 1964.

Marder, Michael. *Dust*. New York: Bloomsbury, 2016.

Marino, Mark C. "The Racial Formation of Chatbots." *CLCWeb: Comparative Literature and Culture* 16, no. 5 (2014): 1–12. https://doi.org/10.7771/1481-4374.2560.

Marte-Wood, Alden Sajor, and Stephanie Dimatulac Santos. "Circuits of Care: Filipino Content Moderation and American Infostructures of Feeling." *Verge: Studies in Global Asias* 7, no. 2 (Fall 2021): 101–27. https://doi.org/10.1353/vrg.2021.0007.

Martin, Aryn, Natasha Myers, and Ana Viseu. "The Politics of Care in Technoscience." *Social Studies of Science* 45, no. 5 (2015): 625–41.

Marx, Karl, and Friedrich Engels. *The Communist Manifesto*. Edited by Gareth Stedman Jones. Translated by Samuel Moore. New York: Penguin Classics, 2002.

Marx, Leo. "'Technology': The Emergence of a Hazardous Concept." *Social Research* 64, no. 3 (1997): 965–88.

Mattern, Shannon. *A City Is Not a Computer: Other Urban Intelligences*. Princeton, NJ: Princeton University Press, 2021.

McElroy, Erin. "Dis/Possessory Data Politics: From Tenant Screening to Anti-eviction Organizing." *International Journal of Urban and Regional Research* 47, no. 1 (2022): 54–70.

McElroy, Erin. "The Work of Landlord Technology: The Fictions of Frictionless Property Management." *Environment and Planning D: Society and Space*, May 8, 2024, https://doi.org/10.1177/02637758241232758.

McElroy, Erin, and Manon Vergerio. "Automating Gentrification: Landlord Technologies and Housing Justice Organizing in New York City Homes." *Environment and Planning D: Society and Space* 40, no. 4 (2022): 607–26.

McGuirk, Justin. "Interview: Blaise Aguera y Arcas." *Surrealism and Design Now*, 198–205.

McKittrick, Katherine. *Dear Science and Other Stories*. Durham, NC: Duke University Press, 2021.

Melamed, Jodi. "Racial Capitalism." *Critical Ethnic Studies* 1, no. 1 (Spring 2015): 76–85.

Mills, Mara. "On Disability and Cybernetics: Helen Keller, Norbert Wiener, and the Hearing Glove." *Differences* 22, nos. 2–3 (2011): 74–111. https://doi.org/10.1215/10407391-1428852.

Minton, Stephen James, and Helene Thiesen. "Greenland." In *Residential Schools and Indigenous Peoples: From Genocide via Education to the Possibilities for Processes of Truth, Restitution, Reconciliation, and Reclamation*, edited by Stephen James Minton, 95–112. London: Routledge, 2019.

Molina, Natalia. *Fit to Be Citizens? Public Health and Race in Los Angeles, 1879–1939*. Berkeley: University of California Press, 2006.

Moran, Taylor C. "Racial Technological Bias and the White, Feminine Voice of AI VAs." *Communication and Critical/Cultural Studies* 18, no. 1 (2021): 19–36. https://doi.org/10.1080/14791420.2020.1820059.

Morozov, Evgeny. "Critique of Techno-feudal Reason." *New Left Review*, nos. 133/134 (2022): 89–126.

Mukogosi, Joan. "Establishing Vigilant Care: Data Infrastructures and the Black Birthing Experience." Data & Society report, July 10, 2024. https://datasociety.net/wp-content /uploads/2024/07/establishing_vigilant_care_report.pdf.

Muñoz Martinez, Monica. *The Injustice Never Leaves You: Anti-Mexican Violence in Texas*. Cambridge, MA: Harvard University Press, 2018.

Murdock, Graham. "Media Materialties: For a Moral Economy of Machines." *Journal of Communication* 68, no. 2 (2018): 359–68. https://doi.org/10.1093/joc/jqx023.

Murphy, Michelle. *Seizing the Means of Reproduction*. Durham, NC: Duke University Press, 2012.

Nakamura, Lisa, and Peter Chow-White, eds. *Race after the Internet*. New York: Routledge, 2012.

National Police Misconduct Reporting Project. *2010 Annual Report*. Washington, DC: Cato Institute, 2010.

Naughton, John. "The Evolution of the Internet: From Military Experiment to General Purpose Technology." *Journal of Cyber Policy* 1, no. 1 (2016): 5–28. https://doi.org/10 .1080/23738871.2016.1157619.

Navarro, Armando. *The Immigration Crisis: Nativism, Armed Vigilantism, and the Rise of a Countervailing Movement*. Lanham, MD: Altamira, 2008.

Navin Brooks, Andrew. "Fugitive Listening: Sounds from the Undercommons." *Theory, Culture and Society* 37, no. 6 (2020): 25–45. https://doi.org/10.1177/0263276420911962.

Nealon, Jeffrey T. *Plant Theory: Biopower and Vegetable Life*. Stanford, CA: Stanford University Press, 2016.

Nelson, Alondra. *Body and Soul*. Minneapolis: University of Minnesota Press, 2011.

Nelson, Alondra. "Introduction: Future Texts." *Social Text* 20, no. 2 (2002): 1–15. https:// muse.jhu.edu/article/31931.

Nelson, Jennifer. "Abortions under Community Control: Feminism, Nationalism, and the Politics of Reproduction among New York City's Young Lords." *Journal of Women's History* 13, no. 1 (2001): 157–80.

Ngai, Mae M. *Impossible Subjects: Illegal Aliens and the Making of Modern America*. 2004. Reprint, Princeton, NJ: Princeton University Press, 2005.

Noble, Safiya Umoja. *Algorithms of Oppression: How Search Engines Reinforce Racism*. New York: NYU Press, 2018.

Nozick, Robert. *Anarchy, State, and Utopia*. New York: John Wiley and Sons, 1974.

Okafor, Chinyere. "Black Feminism Embodiment: A Theoretical Geography of Home, Healing, and Activism." *Meridians* 16, no. 2 (2018): 373–81.

O'Mara, Margaret. *The Code: Silicon Valley and the Remaking of America*. New York: Penguin, 2020.

O'Neil, Cathy. *Weapons of Math Destruction: How Big Data Increases Inequality and Threatens Democracy*. New York: Crown, 2016.

Ong, Aihwa. *Neoliberalism as Exception: Mutations in Citizenship and Sovereignty*. Durham, NC: Duke University Press, 2006.

Osucha, Eden. "The Whiteness of Privacy: Race, Media, Law." *Camera Obscura* 24, no. 1(70) (2009): 67. https://doi.org/10.1215/02705346-2008-015.

Oyama, Rebecca. "Do Not (Re)Enter: The Rise of Criminal Background Tenant Screening as a Violation of the Fair Housing Act." *Michigan Journal of Race and Law* 15, no. 1 (2009): 181–222.

Paasonen, Susanna. 2018. *Many Splendored Things: Thinking Sex and Play*. London: Goldsmiths Press.

Parvin, Nassim. "Look Up and Smile! Seeing through Alexa's Algorithmic Gaze." *Catalyst: Feminism, Theory, Technoscience* 5, no. 1 (2019): 1–11. https://doi.org/10.28968/cftt.v5i1.29592.

Petryna, Adriana. *When Experiments Travel: Clinical Trials and the Global Search for Human Subjects*. Princeton, NJ: Princeton University Press, 2009.

Phan, Thao. "Amazon Echo and the Aesthetics of Whiteness." *Catalyst: Feminism, Theory, Technoscience* 5, no. 1 (2019): 1–38. https://doi.org/10.28968/cftt.v5i1.29586.

Phipps, Alison. "'Every Woman Knows a Weinstein': Political Whiteness and White Woundedness in #MeToo and Public Feminisms around Sexual Violence." *Feminist Formations* 31, no. 2 (2019): 1–25.

Pickens, Therí Alyce. *Black Madness:: Mad Blackness*. Durham, NC: Duke University Press, 2019.

Piepzna-Samarasinha, Leah Lakshmi. *Care Work: Dreaming Disability Justice*. Vancouver, BC: Arsenal Pulp, 2018.

Povinelli. Elizabeth A. *Geontologies: A Requiem to Late Liberalism*. Durham, NC: Duke University Press, 2016.

Power, Emma R., and Charles Gillon. "Performing the 'Good Tenant.'" *Housing Studies* 37, no. 3 (2022): 459–82. https://doi.org/10.1080/02673037.2020.1813260.

Prabhu, Vinay Uday, and Abeba Birhane. "Large Image Datasets: A Pyrrhic Win for Computer Vision?" In *Proceedings of 2021 IEEE Winter Conference on Applications of Computer Vision*, Ithaca, NY: Cornell University Press, 2021: 1536–46.

Pradhan, Alisha, Leah Findlater, and Amanda Lazar. "'Phantom Friend' or 'Just a Box with Information': Personification and Ontological Categorization of Smart Speaker-Based Voice Assistants by Older Adults." *Proceedings of the ACM on Human-Computer Interaction* 3, no. 214 (2019): 1–21.

Pradhan, Alisha, Kanika Mehta, and Leah Findlater. "'Accessibility Came by Accident': Use of Voice-Controlled Intelligent Personal Assistants by People with Disabilities." In *Proceedings of the 2018 CHI Conference on Human Factors in Computing Systems*, 1–13. Montreal QC, Canada, 2018. https://doi.org/10.1145/3173574.3174033.

Purvis, Dara E., and Melissa Blanco. "Police Sexual Violence: Police Brutality, #MeToo, and Masculinities." *California Law Review*, vol. 198 no. 5 (2020): 1487–529.

Raley, Rita. "Border Hacks: Electronic Civil Disobedience and the Politics of Immigration." In *Tactical Media*, 31–64. Minneapolis: University of Minnesota Press, 2009.

Ralph, Michael. "'Life . . . in the Midst of Death': Notes on the Relationship between Slave Insurance, Life Insurance, and Disability." *Disability Studies Quarterly* 32, no. 3 (2012), https://dsq-sds.org/index.php/dsq/article/view/3267.

Ramadan, Zahy, Maya F. Farah, and Lea El Essrawi. "From Amazon.com to Amazon.love: How Alexa Is Redefining Companionship and Interdependence for People with Special Needs." *Psychology and Marketing* 38, no. 4 (2021): 596–609. https://doi.org/10.1002/mar.21441.

Reay, Donald T., Corinne L. Fligner, Allan D. Stilwell, and Judy Arnold. "Positional Asphyxia during Law Enforcement Transport." *American Journal of Forensic Medicine and Pathology* 13, no. 2 (1992): 90–97.

Rhee, Jennifer. *The Robotic Imaginary: The Human and the Price of Dehumanized Labor.* Minneapolis: University of Minnesota Press, 2018.

Ribes, David, and Steven Jackson. "Data Bite Man: The Work of Sustaining a Long-Term Study." In Gitelman, *"Raw Data" Is an Oxymoron*, 147–66.

Ritchie, Andrea J. *Invisible No More: Police Violence against Black Women and Women of Color.* Boston: Beacon, 2017.

Rivera, Isaac. "Undoing Settler Imaginaries: (Re)Imagining Digital Knowledge Politics." *Progress in Human Geography* 47, no. 2 (2023): 298–316.

Roberts, Sarah T. "Your AI Is Human." In *Your Computer Is on Fire*, edited by Thomas S. Mullaney, Benjamin Peters, Mar Hicks, and Kavita Philip, 51–71. Cambridge, MA: MIT Press, 2021.

Robinson, Cedric. *Black Marxism: The Making of the Black Radical Tradition.* Chapel Hill: University of North Carolina Press, 1983.

Robinson, Dylan. *Hungry Listening: Resonant Theories for Indigenous Sound Studies.* Minneapolis: University of Minnesota Press, 2020.

Roh, David S., Betsy Huang, and Greta A. Niu, eds. *Techno-Orientalism: Imagining Asia in Speculative Fiction, History, and Media.* New Brunswick, NJ: Rutgers University Press, 2015.

Rojas, David. "Disjointed Times in 'Climate-Smart' Amazonia." *Environmental Humanities* 14, no. 2 (2022): 321–40. https://doi.org/10.1215/22011919-9712401.

Roosth, Sophia. "Screaming Yeast: Sonocytology, Cytoplasmic Milieus, and Cellular Subjectivities." *Critical Inquiry* 35, no. 2 (2009): 332–50. https://doi.org/10.1086/596646.

Rosa, Jonathan, and Nelson Flores. "Unsettling Race and Language: Toward a Raciolinguistic Perspective." *Language in Society* 46, no. 5 (2017): 621–47. https://doi.org/10.1017/S0047404517000562.

Roy, Ananya. "Dis/Possessive Collectivism: Property and Personhood at City's End." *Geoforum*, no. 80 (2017): 1–11.

Roy, Ananya. "Undoing Property: Feminist Struggle in the Time of Abolition." *Society + Space.* May 3, 2021. societyandspace.org/articles/undoing-propert-feminist-struggle-in-the-time-of-abolition.

Roy, Deboleena. *Molecular Feminisms: Biology, Becomings, and Life in the Lab.* Seattle: University of Washington Press, 2018.

Sadowski, Jathan, Yolande Strengers, and Jenny Kennedy. "More Work for Big Mother: Revaluing Care and Control in Smart Homes." *Environment and Planning A: Economy and Space* 56, no. 1 (2021): 330–45. https://doi.org/10.1177/0308518X211022366.

Scalisi, Joseph Frank, Gregory Saul Harrison, Desiree Mejia, and Andrew Paul Thomas. Doorbell Chime Systems and Methods. US Patent 9,179,107 B1, filed May 28, 2015, and issued November 3, 2015. https://patents.google.com/patent/US9179107B1/en.

Scalisi, Joseph Frank, and Seton Paul Kasmir. Doorbell Communication Systems and Method. U.S. Patent no. 9,055,202 B1, filed February 12, 2015, and issued June 9, 2015. https://patents.google.com/patent/US9055202B1/en.

Schüll, Natasha Dow. "Data for Life: Wearable Technology and the Design of Self-Care." *BioSocieties*, vol. 11, no. 3 (2016): 317–33.

Schultz-Figueroa, Benjamin. "Glitch/Glitsch: (More Power) Lucky Break and the Position of Modern Technology." *Culture Machine* 12 (2011), 9.

Seberger, John S., Irina Shklovski, Emily Swiatek, and Sameer Patil. "Still Creepy after All These Years: The Normalization of Affective Discomfort in App Use." In *Proceedings of the 2022 CHI Conference on Human Factors in Computing Systems*, no. 159 (2022): 1–19.

Semel, Beth M. "Listening like a Computer: Attentional Tensions and Mechanized Care in Psychiatric Digital Phenotyping." *Science, Technology, and Human Values* 47, no. 2 (2022): 266–90. https://doi.org/10.1177/01622439211026371.

Semel, Beth M. "Speech, Signal, Symptom: Machine Listening and the Remaking of Psychiatric Assessment." PhD diss., Massachusetts Institute of Technology, 2019.

Shah, Nayan. *Contagious Divides: Epidemics and Race in San Francisco's Chinatown.* Berkeley: University of California Press, 2001.

Sharma, Sarah. "A Manifesto for the Broken Machine." *Camera Obscura* 35, no. 2(104) (2020): 171–79. https://doi.org/10.1215/02705346–8359652.

Shelby, R., Jenna I. Harb, and Kathryn E. Henne. "Whiteness in and through Data Protection: An Intersectional Approach to Anti-violence Apps and #MeToo Bots." *Internet Policy Review* 10, no. 4 (2021): 1–25.

Silverstein, Michael. "Translation, Transduction, Transformation: Skating 'Glossando' on Thin Semiotic Ice." In *Translating Cultures: Perspectives on Translation and Anthropology*, edited by Paula G. Rubel and Abraham Rosman, 75–100. Berg: Oxford University Press, 2003.

Sim, Kate. "Respond and Resolve: A Critical Feminist Inquiry for Technologies of Sexual Governance." Global Perspectives 2, no. 1 (2021): 254–334. https://doi.org/10.1525/gp .2021.25434.

Siminoff, James, August Cziment, Aaron Harpole, Elliott Lemberger, John Modestine, and Darrell Sommerlatt. Sharing video footage from audio/video recording and communication devices. U.S. Patent 9,819,713 B2, filed Febuary 13, 2017, and issued November 14, 2017. https://patents.google.com/patent/US10033780B2/en.

Simon, Phil. "Racial Profiling at Nextdoor: Using Data to Build a Better App and Combat a PR Disaster." In *Analytics: The Agile Way*, 193–220. Newark, NJ: John Wiley and Sons, 2017.

Slatton, Brittany C., and April L. Richard. "Black Women's Experiences of Sexual Assault and Disclosure: Insights from the Margins." *Sociology Compass* 14, no. 6 (2020): e12792.

Smith, Jason E. *Smart Machines and Service Work: Automation in an Age of Stagnation.* London: Reaktion Books, 2020.

Spencer, Glenn, and Michael S. Davis. Barrier detection system and method. US Patent 9,625,594 B2, filed April 18, 2017.

Stallabrass, Julian. "Sublime Calculation." *New Left Review*, no. 132 (November/December 2021). https://newleftreview.org/issues/ii132/articles/julian-stallabrass-sublime -calculation.

Star, Susan Leigh. "The Ethnography of Infrastructure." *American Behavioral Scientist* 43, no. 3 (1999): 377–91. https://doi.org/10.1177/00027649921955326.

Stark, Luke, and Kate Crawford. "The Work of Art in the Age of Artificial Intelligence: What Artists Can Teach Us about the Ethics of Data Practice." *Surveillance and Society* 17, no. 3/4 (2019): 442–55.

Stauffer, Robert R. "Tenant Blacklisting: Tenant Screening Services and the Right to Privacy Note." *Harvard Journal on Legislation* 24, no. 1 (1987): 239–314.

Steinfeld, Aaron, Chadwicke Jenkins Odest, and Brian Scassellati. "The Oz of Wizard: Simulating the Human for Interaction Research." In *Proceedings of the 4th ACM/IEEE International Conference on Human Robot Interaction*, 101–8. La Jolla, CA: ACM Press, 2009. https://doi.org/10.1145/1514095.1514115.

Stern, Alexandra Minna. "Buildings, Boundaries, and Blood: Medicalization and Nation-Building on the US-Mexico Border, 1910–1930." *Hispanic American Historical Review* 79, no. 1 (1999): 41–81.

Stern, Alexandra Minna. *Eugenic Nation: Fault and Frontiers of Better Breeding in Modern America*. Berkeley: University of California Press, 2005.

Stern, Alexandra Minna. *Proud Boys and the White Ethno-state: How the Alt-Right Is Warping the Imagination*. Boston: Beacon, 2019.

Sterne, Jonathan. *The Audible Past*. Durham, NC: Duke University Press, 2003.

Stewart, Whitney Nell. "The Racialized Politics of Home in Slavery and Freedom." PhD diss., Rice University, 2017.

Stoler, Ann Laura. "Intimidations of Empire: Predicaments of the Tactile and Unseen." In *Haunted by Empire: Geographies of Intimacy in North American History*, edited by Ann Laura Stoler, 1–22. Durham, NC: Duke University Press, 2006. https://doi.org/10.1515/9780822387992-003.

Strengers, Yolande, and Jenny Kennedy. *The Smart Wife*. Cambridge, MA: MIT Press, 2020.

Summers, Brandi T., and Desiree Fields. "Speculative Urban Worldmaking: Meeting Financial Violence with a Politics of Collective Care." *Antipode*, 56, no. 3 (2022): 821–40. https://doi.org/10.1111/anti.12900.

Terry, Jennifer. *Attachments to War: Biomedical Logics and Violence in Twenty-First-Century America*. Durham, NC: Duke University Press, 2017.

Thakor, Mitali. "Deception by Design: Digital Skin, Racial Matter, and the New Policing of Child Sexual Exploitation." In Benjamin, *Captivating Technology*, 188–208.

Thiesen, Helene. *Greenland's Stolen Indigenous Children*. Translated by Stephen Minton. London: Routledge, 2022.

Thylstrup, Nanna Bonde, Daniela Agostinho, Annie Ring, Catherine D'Ignazio, and Kristin Veel. *Uncertain Archives: Critical Keywords for Big Data*. Cambridge, MA: MIT Press, 2021.

Tsing, Anna Lowenhaupt. *The Mushroom at the End of the World: On the Possibility of Life in Capitalist Ruins*. Princeton, NJ: Princeton University Press, 2015.

Tuck, Eve K., and Wayne Yang. "Decolonization Is Not a Metaphor." *Decolonization: Indigineity, Education, and Society* 1, no. 1 (2012): 1–40.

Tuerkheimer, Deborah. "Incredible Women: Sexual Violence and the Credibility Discount." *University of Pennsylvania Law Review* 166, no. 1 (2017): 1–58.

Ukeles, Mierle Laderman. "Manifesto! Maintenance Art—Proposal for an exhibition: 'CARE,'" Philadelphia, PA, October 1969. https://queensmuseum.org/wp-content

/uploads/2016/04/Ukeles-Manifesto-for-Maintenance-Art-1969.pdf Accessed July 31, 2024.

Uscinski, Joseph E., and Joseph M. Parent. *American Conspiracy Theories*. Oxford: Oxford University Press, 2014.

US Government Accountability Office. *Border Patrol: Key Elements of New Strategic Plan Not Yet in Place to Inform Border Security Needs and Resource Needs*. GAO-13-25. Washington, DC: Government Accountability Office, 2012. www.gao.gov/products /gao-13-25.

van Dijck, Jose. "Datafiction, Dataism and Dataveillance: Big Data between Scientific Paradigm and Secular Belief." *Surveillance and Society* 12, no. 2 (2014): 197–208.

Vasquez-Tokos, Jessica, and Priscilla Yamin. "The Racialization of Privacy: Racial Formation as a Family Affair." *Theory and Society* 50, no. 5 (2021), 717–40.

Venema, Rachel M., Katherine Lorenz, and Nicole Sweda. "Unfounded, Cleared, or Cleared by Exceptional Means: Sexual Assault Case Outcomes from 1999 to 2014." *Journal of Interpersonal Violence* 36 no. 19–20 (2019): 0886260519876718.

Vint, Sherryl. *Science Fiction: A Guide for the Perplexed*. London: Bloomsbury, 2014.

Vorsino, Zoe. "Chatbots, Gender, and Race on Web 2.0 Platforms: Tay. AI as Monstrous Femininity and Abject Whiteness." *Signs* 47, no. 1 (2021): 105–27.

Walcott, Rinaldo. *On Property*. Windsor, ON: Biblioasis, 2021.

Walker, Jesse. *The United States of Paranoia: A Conspiracy Theory*. New York: Harper Perennial, 2014.

Walker, Richard. *Pictures of a Gone City: Tech and the Dark Side of Prosperity in the San Francisco Bay Area*. Oakland, CA: PM Press, 2018.

Warren, Samuel, and Louis D. Brandeis. "The Right to Privacy." *Harvard Law Review* 4, no. 5 (1890): 193–220.

Weheliye, Alexander G. "'Feenin': Posthuman Voices in Contemporary Black Popular Music." *Social Text* 20, no. 2 (2002): 21–47. https://doi.org/10.1215/01642472-20-2 _71-21.

Weinbaum, Alys Eve. *The Afterlife of Reproductive Slavery: Biocapitalism and Black Feminism's Philosophy of History*. Durham, NC: Duke University Press, 2019.

Weitzel, Michelle D. "Audializing Migrant Bodies: Sound and Security at the Border." *Security Dialogue* 49, no. 6 (2018): 421–37. https://doi.org/10.1177/0967010618795788.

West, Emily. "Amazon: Surveillance as a Service." *Surveillance and Society* 17, no. 1/2 (2019): Platform Surveillance 27–33. https://doi.org/10.24908/ss.v17i1/2.13008.

West, Sarah Myers. "Data Capitalism: Redefining the Logics of Surveillance and Privacy." *Business and Society* 58, no. 1 (2019): 20–41.

Wexler, Laura. *Tender Violence: Domestic Visions in an Age of US Imperialism*. Chapel Hill: University of North Carolina Press, 2000.

Whyte, Kyle Powys. "Indigenous Women, Climate Change Impacts, and Collective Action." *Hypatia* 29, no. 3 (2014): 599–616. www.jstor.org/stable/24542019.

Wiehn, Tanja Anna. "Algorithmic Intimacies: A Cultural Analysis of Ubiquitous Proximities in Data." PhD diss., University of Copenhagen, 2021.

Winner, Langdon. *Autonomous Technology: Technics-out-of-Control as a Theme in Political Thought*. Cambridge, MA: MIT Press, 2001.

Wood, Catherine. *Yvonne Rainer: The Mind Is a Muscle*. London: Afterall, 2007.

Woods, Heather Suzanne. "Asking More of Siri and Alexa: Feminine Persona in Service of Surveillance Capitalism." *Critical Studies in Media Communication* 35, no. 4 (2018): 334–49. https://doi.org/10.1080/15295036.2018.1488082.

Wosk, Julie. *My Fair Ladies: Female Robots, Androids, and Other Artificial Eves.* New Brunswick, NJ: Rutgers University Press, 2015.

Woźniak, Paweł W., Jakob Karolus, Florian Lang, Caroline Eckerth, Johannes Schöning, Yvonne Rogers, and Jasmin Niess. "Creepy Technology: What Is It and How Do You Measure It?" *CHI '21: Proceedings of the 2021 CHI Conference on Human Factors in Computing Systems*, no. 719, 1–13. 2021. https://dl.acm.org/doi/10.1145/3411764.3445299.

Zelizer, Viviana A. Rotman. *The Purchase of Intimacy.* Princeton, NJ: Princeton University Press, 2005.

Zitzewitz, Karin, and Mithu Sen. "'We Have Entered a Stage of Overproduction': The Weight of (Un)Hospitality and the Performative Self." *Critical Times* 4, no. 1 (2021): 174–85.

Zuboff, Shoshana. *The Age of Surveillance Capitalism.* London: Profile Books, 2019.

Zuijderduijn, Jaco, and Roos van Oosten. "Breaking the Piggy Bank: What Can Historical and Archaeological Sources Tell Us about Late-Medieval Saving Behaviour?" Centre for Global Economic History Working Paper Series No. 65, Utretcht University, Utretcht, Netherlands, June 2015. https://econpapers.repec.org/paper/ucgwpaper/0065.htm.

Contributors

NEDA ATANASOSKI is professor and chair of the Harriet Tubman Department of Women, Gender and Sexuality Studies at the University of Maryland, College Park, and associate director of the Artificial Intelligence Interdisciplinary Institute at Maryland (AIM). She is the author of *Humanitarian Violence: The US Deployment of Diversity* (2013), coauthor of *Surrogate Humanity: Race, Robots and the Politics of Technological Futures* (2019), and coeditor of *Postsocialist Politics and the Ends of Revolution* (2022).

KATHERINE BENNETT is a PhD student in digital media at the Georgia Institute of Technology. They research aesthetics of Black feminist and queer technoculture, in particular the representational landscapes of Afrofuturist literature and visual media. Their writing and low-tech artifacts and installations examine connections between art and nature. Katherine has practiced and taught landscape architecture in the United States and South Korea.

IVÁN CHAAR LÓPEZ is assistant professor in the Department of American Studies at the University of Texas at Austin and principal investigator of the Border Tech Lab. He is the author of *The Cybernetic Border: Drones, Technology, and Intrusion* (2024) and a coauthor of Precarity Lab's *Technoprecarious* (2020).

SUSHMITA CHATTERJEE is chair and professor of ethnic studies and women's and gender studies at Colorado State University. Sushmita's research interests include postcolonial studies, animal studies, feminist-queer theory, and visual politics. Her book *Postcolonial Hauntings: Play and Transnational Feminism* was published by the University of Illinois Press in 2024.

HAYRI DORTDIVANLIOGLU is a PhD candidate and a Fulbright Scholar in the School of Architecture at the Georgia Institute of Technology. Hayri's scholarly interests include architectural history, theory, design, critical computation, representation, craft, and fabrication. He utilizes speculative mapping and data visualization to question the interaction between humans, their environments, and technology.

JACOB HAGELBERG received his PhD in Cultural Studies from the University of California, Davis, with designated emphases in Science and Technology Studies and Critical Theory. He is working on his first book project, *Techlash Media: Credit, Credibility, and Crisis*, which engages the cultural politics of late techno-capitalism and its discontents. He has also taught computer science ethics, a media theory course on *Black Mirror*, and is currently Guest Faculty at the University of Washington's Information School.

SANAZ HAGHANI, an interdisciplinary artist and faculty member at Rowan-Cabarrus Community College, specializes in installations, art books, and handmade paper that delve into themes of womanhood and cultural identity. Holding a master of fine arts degree from the University of Georgia, her art has been featured in significant exhibitions, most notably *Paper Routes: Women to Watch 2020* at the Museum of Contemporary Art of Georgia.

JENNIFER A. HAMILTON is professor and chair of anthropology at Bates College. She is a sociocultural anthropologist whose interdisciplinary research and teaching focus on feminist science and technology studies, medical and legal anthropology, ethnography, and the politics of indigeneity. She is the author of *Indigeneity in the Courtroom* (2009) and is currently revising her second book manuscript, *Settler Science and the Politics of Indigeneity*.

ANTONIA HERNÁNDEZ is an artist and assistant professor in the Department of Communication Studies at Concordia University. She combines art practice with theoretical investigation to explore the poetics of governance and the infrastructural dimension of intimacy. She has presented her work and creative methods extensively at conferences, exhibitions and art talks. Her current project, *Hydrofictions*, investigates imaginaries about water, including financial and speculative ones.

MARJAN KHATIBI is assistant professor in the Department of Design at San Jose State University. She holds an MFA in digital arts and new media from the University of California, Santa Cruz. She is an interdisciplinary visual designer from Tehran, Iran. She specializes in virtual reality and augmented reality and design for social change. She has presented research at conferences such as the AIGA Design Educators Community, the Broadcast Education Association, and the University and College Designers Association.

TAMARA KNEESE is Director of the Climate, Technology, and Justice Program at the Data and Society Research Institute. She is the author of *Death Glitch: How Techno-solutionism Fails Us in This Life and Beyond* (2023) and a coeditor of *The New Death: Mortality and Death Care in the Twenty-First Century* (2022).

ERIN MCELROY is assistant professor in the Department of Geography at the University of Washington, where they run the Anti-Eviction Lab, and is cofounder of the Anti-Eviction Mapping Project and the *Radical Housing Journal*. Erin is the author of *Silicon Valley Imperialism: Techno Fantasies and Frictions in Postsocialist Times* (2024) and coeditor of *Counterpoints: A San Francisco Bay Area Atlas of Displacement and Resistance* (2021).

VERNELLE A. A. NOEL, PhD, is a computational design scholar, architect, and artist and director of the Situated Computation and Design Lab. Using interdisciplinary approaches, she builds new frameworks, methodologies, and tools to explore social, cultural, and political aspects of computation and emerging technologies for new reconfigurations of practice, pedagogy, and publics. Areas include craft, vernacular, and digital practices and automated and emerging technologies.

JESSICA L. OLIVARES is a postdoctoral fellow in the Department of Bioethics at the University of Texas Medical Branch. In 2023, she successfully defended her doctoral dissertation at Rice University on the surveillance state. She received a joint MA in anthropology and women's, gender, and sexualities studies at Brandeis University.

NASSIM PARVIN is associate professor at the University of Washington Information School, where she also serves as associate dean for inclusion, diversity, equity, access and sovereignty. Dr. Parvin's interdisciplinary research integrates theoretically driven humanistic scholarship and design-based inquiry. Her papers have appeared in design, Human Computer Interaction (HCI), and Science and Technology Studies (STS) venues. Her designs have been deployed at nonprofit organizations and exhibited in venues such as the Smithsonian Museum. She is an award-winning educator and one of the lead coeditors of *Catalyst: Feminism, Theory, Technoscience*.

BETH SEMEL is assistant professor in the Department of Anthropology at Princeton University. Drawing from linguistic and medical anthropology, and feminist science and technology studies, her research explores the technopolitics and sensory politics of machine listening in American mental health care.

RENEE SHELBY is a visiting fellow at the Justice and Technoscience Lab in the School of Regulation and Governance at Australian National University. She researches and writes about the social impacts of technology, with a focus on how to develop technologies that promote equity. Her approach is informed by her background as a public sociologist working with antiviolence and LGBTQ organizations in the US South.

TANJA WIEHN is an Assistant Professor at the Department of Arts and Cultural Sciences at Lund University. In 2021, she received her PhD from the University of Copenhagen with her thesis on algorithmic intimacies. In her research, she investigates algorithmic systems, digital culture, and artistic (research) practices through feminist epistemologies.

Index

American Border Patrol (ABP), 18, 50–52, 54–61
American Civil Liberties Union(ACLU), 61
American Medical Association, 177
Amoore, Louise, 13, 118
anachronic time, 238
Andalibi, Nazanin, 183
Andrejevic, Mark, 112
Anduril Industries, 59
animals, 4, 22, 176, 182, 238–49; stuffed, 1, 97, 166, 222
Anishinaabe people, 234–35
Ant Financial, 195
anthropocentrism, 238, 244–45
anthropology, 54, 210
anthropomorphism, 78, 150, 245
anti-Blackness, 128. *See also* racism
Anti-Eviction Lab, 81, 90
Anti-Eviction Mapping Project, 80–81, 89–90
anti-immigration politics, 50, 54–57, 61. *See also* immigration
antirape technology. *See* rape-reporting apps
antiviolence technology, 66–67, 73. *See also* rape-reporting apps
anxiety, 1, 14, 21, 68, 103, 183, 188, 190; over privacy, 98, 107
Anzaldúa, Gloria, 144n34
Apple, 11–*12*, 83, 107, 179
Aristotle, 244
Arpaio, Joe, 60
artifacts, 191, 217, 244
artificial intelligence (AI), xv–xvi, 3, 6, 241, 248, 260; art generated by, 255–59; and desire, 161, 165–68; in landlord technologies, 18, 79–81, 86; and rape-reporting apps, 18, 67, 69, 71–73; of smart home assistants, 20, 111, 114, 117–19, 121, 148, 150–51, 155–56, 241; in social credit system, 188, 190, 198; and surveillance, 9, 13–14; in transductive labor, 206–8, 212–14. *See also* Alexa; Chat-GPT; Cherry Home; Google, Deep Dream Generator; Talk to Spot
artwork, 22, 40, 109, 160; AI, 255; and nature, 233, 240–41; performance, 19, 21, 111–13, 119–22; of smart cities, 230–31
Asia, 120–21, 188, 190, 200. *See also* China; India
Asian Americans, 86, 120

assistive technology, 3. *See also* smart home assistants
Atanasoski, Neda, 51, 53, 79, 119, 150, 174, 248
Atlanta, Georgia, 8, 149, 231
audiences, 28, 33, 35–36, 46, 58, 191, 196, 241–42
augmentation, 4, 6, 42, 55, 83, 98, 102–3, 150, 224, 245; of reality, 164, 192
Aunt Jemima, 103
Australia, 175, 241; Bawaka Country, 233–34
authoritarianism, 21, 197, 199–200
automated landlords, 85. *See also* landlord technologies
automation, 15–16, 46, 50, 119–21, 152, 173, 197, 258–59; and hometech, 19, 97, 101–2, 107–8, 112, 114; in landlord technologies, 80, 83, 85–86, 89; and transductive labor, 21, 205, 209–12, 214, 216–18
automatons, 120–21, 164
avatars, xvi, 9, 166, 205, 208, 212–17

BabyCenter (app), 176
Baldwin, James: "Evidence of Things Not Seen, The," 7–8
Ballard, J. G., 94, 109
Balsamo, Anne, 69
Barthes, Roland, 32
Beck, Alex, 259
Been, Vicky, 86
Behar, Katherine: *Pipecleaner*, 45–46
Beller, Jonathan, 63n43
Benjamin, Ruha, 73, 90
Bennett, Katherine: *Street Smarts*, 22, 221–22, 230–31
Benson, Robert, 83
Berlant, Lauren, 82
Bezos, Jeff, 149
Bhandar, Brenna, 79
biases, 72, 83, 119, 136, 245, 260; algorithmic, 13, 78, 115, 227; racial, 87, 121, 259
Biering, Raymond, 83
Big Brother, 197
big data, 115, 148, 188, 197, 222
big tech, 120–21, 190–91, 199
binaries, x, 3, 11, 179, 209, 211, 217, 226; surveillance/privacy, 16, 107, 113, 174, 207
Bing, 165–66
BioConnect, 88
biomarkers, vocal, 21, 206–16, 218

creep, sensing, 4, 7, 9, 14, 16, 22–23, 27–28, 160, 176, 197–98; in home surveillance, 126; and nationalism, 200; of sexcam platforms, 34, 38; and smart desires, 164–68; of smart dust, 41; and smart homes, 98–100, 108; and uncanny valley, 166–67; and vocal biomarker technologies, 207–8

creep, slow movement of, 4–6, 8, 16, 21–23, 51, 198–99, 244, 249, 255; and border politics, 55–56, 59–60; in COVID-19 times, 239, 248–50; in home surveillance, 127; in home technologies, 18–19, 99, 104, 108, 115; and smart forests, 226–27

creep, technological (technocreep): and citizen science, 51, 57–58; in COVID-19 times, 238–41, 245–50; of home security, 135; in landlord technologies, 79, 90; and maintenance play, 29; meaning of, xv–xvi, 4–11, 16–23; and nationalism, 198–99; and Porkfolio, 104, 107; and rape-reporting apps, 66–69, 72; in science fiction, 191–92; and self-tracking apps, 174, 176, 181, 183; and smart dust, 44, 46; of smart forests, 224; of smart home assistants, 111–13, 121, 148–49, 153, 155–56; study of, 1, 3

creep, intuition of, 4–5, 7, 16, 21–22, 183, 190–92. *See also* creep, sensing

creep, sensing, 4, 21–22, 166, 168, 198, 23n7

criminalization, 68, 177, 182

critical race science and technology studies, 12, 155, 209

crowdsourcing, 19, 55, 173, 176–77, 182–83

Crowley, Amanda McDonald, *113*

cruel optimism, 82

cybernetics, 57, 59, 83, 210

cyberpunk, 188

DALL-E, 258–59

dark sousveillance, 24n30

DARPA, 43

data, 10, 78, 161, 197, 231, 246, 258; -based surveillance, 174–75; bias, 259–60; big, 115, 148, 188, 197, 222; in border technology, 50–52, 57–60; capitalism, 116, 119, 121–22, 123n37; in citizen science, 53; in citizen scoring, 188, 195–96, 198; commodification of, 3, 9, 82, 178; health, 178–80, 182, 184; in landlord technology, 80–83, 85–86, 88–90; and machine vision, 116–18, 208; monetization of, 17, 107, 128,

179; personal, xv, 3, 11, 179, 182, 207; privacy, 10–11, 107, 113, 177, 191; as property, 107–8; on rape-reporting apps, 18, 67, 70–72; and smart dust, 41–42, 44, 46; from smart forests, 21, 223–25, 227; and smart home assistants, 17, 112, 115–16, 121, 148, 150–51, 156; and smart homes, 3, 9, 99, 102, 126; on social media, xv, 11, 176; and technocapitalism, 22, 79; in transductive labor, 205–16

datafication, 1, 19, 83, 86, 111–15, 117, 119, 121

dataveillance, 196

dating apps, 20, 161–62

Daum, Jeremy, 191, 197, 202n13

d/Deaf people, 47, 210

decolonialism, 80. *See also* colonialism/ imperialism

deepfakes, 3

deep learning, 117–18, 123n24

Deer, Sarah, 68

defamiliarization, 27, 115

deforestation, 223–25

dehumanization, 120–21

Del-Em, 181

Deleuze, Gilles, 230

De León, Jason, 63n49

Deloria, Vine, Jr., 6

demarcations, 5, 79, 140, 238–40; of neighborhoods, 135, 137; of otherness, 126, 128

democracy, 10, 20, 44, 59, 113, 188, 199; and citizen science, 18, 51–52, 55, 61

Deng Xiaoping, 202n17

Denmark, 122n9

dependency, xv–xvi, 42, 103, 105, 115–16, 168, 190, 235, 244; co-, 5, 23, 235; inter-, 11, 148, 154, 156; on maintenance, 226; mutual, 29; on transduction, 210, 217

deportation, 59, 70

depression, 68, 175, 205–6, 212

Derrida, Jacques, 238, 245–46

design, 1, 7, 11, 13, 15, 89, 121, 161, 192, 258; of antirape apps, 67, 69–72; of citizen technoscience, 55–57, 59; and creepiness, xvi, 1, 9; gamified, 195; and people with disabilities, 148, 151–52, 155; persuasive, 162, 170; of Porkfolio, 104, 106; and power relations, 8, 18, 66, 127; of pregnancy-tracking apps, 173, 175–76, 181, 183; sexism in, 119; of smart dust, 44, 46; of vocal biomarker technologies, 205, 207, 211, 214, 216

Design Museum (London), 257

desire, 16, 22, 37, 45, 67, 85, 112, 160, 183; and AI-generated art, 256, 258; and home technologies, 99, 102, 108–9, 126, 130, 138; smart, 17, 20, 161–70; and smart home assistants, 19, 148, 151–52; for technofixes, 20, 149; in transductive labor, 215

detachment, 214, 217, 227

determinism, 112, 118, 121, 210

deviance, 209–10

diagnosis of disenchantment, 116–17

diet apps, 20, 161–62

digital domesticity, 150, 156

digital doormen, 86

digitization, 83–84

Dinkins, Stephanie, 259

disabilities, 167, 179, 206–7, 209, 216, 218; technologies for people with, 19–20, 148–56

disability studies, 19, 209, 217

disaster capitalism, 78, 84

disclosure, 27–28, 32, 34–35, 176, 197; in rape-reporting apps, 66–70, 72–73

displacement, 21, 51, 55, 190, 196, 248

dispossession, 18, 79–80, 82, 84–85, 88, 103, 217

dis/possessive collectivism, 89

dissemination, 52, 54, 150, 177

DMello, Alvin, 258

Dole, Elizabeth, 56

dollhouses, 31–38

domesticity, 15, 103, 130, 149–50; of Alexa, 151, 154–56; creep of, 17, 27, 79; and maintenance work, 45, 153; of nation, 50, 55–56; on sex webcam platforms, 28–29, 31, 36; and surveillance, 78, 80, 86

Dortdivanlioglu, Hayri: *Thousand Dreams of Yamur*, 22, *93*–95, 109

dot-com boom, 79, 84, 86

dreams, ix–x, 138, 150, 199, 226; Google Deep Dream, *253*, 255–58; technodeterministic, 112

drones, ix, xv, 10, 56, 58, 60, 108

Dubrofsky, Rachel E., 12

dust. *See* smart dust

dystopia, 83, 107, 187, 196, 199–200

e-carcerality, 80

Echo, 3, 147, 149–51, 153, 219n23. *See also* Alexa

e-commerce, 195, 240

eldercare industry, 13–14, 102

Electronic Frontier Foundation, 177

elimination, 43, 51, 69–70, 94, 150

ELIZA effect, 123n36

Elwood, Sarah, 90

Elyachar, Julia, 211

Embodied Self, The (Khatibi), 22, *39*–40

empowerment, 7, 78, 120, 190; of rape-reporting apps, 67, 69–70, 73

enchantment, 117–18

energy efficiency, 47, 99, 102

Engel, Friedrich, 88

engineering, 1, 9, 58, 98, 205, 211–12, 216, 222, 245, 257; inequity, 115

environment, 16–17, 42, 58, 61, 87, 95, 162, 224–25, 233, 256; built, 151, 231; and citizen science, 52, 54; domestic, 28, 113; and smart dust, 44, 47; and sustainability, 52, 223

Esquire, 189

ethics, 5, 9–11, 13, 42, 106, 118, 120, 148, 232, 234; of human subject research, 215; and smart home assistants, 116; of smart homes, 98

ethnography, x, 21, 126, 205, 207–8, 210, 212–14, 227

eugenics, 86, 107, 152, 177

Eve (artificial), 164–65

evictions, 9, 78, 80–82, 85–86, 88–90

Evictorbook, 89

"Evidence of Things Not Seen, The" (Baldwin), 7–8

e-waste, 8, 46, 224

exclusion, 13, 50, 127, 238, 244–45; and citizen science, 52–55, 59–60; in property, 81–82

Expectful (app), 178

experience machine, 163

exploitation, 8, 22, 35, 103, 151, 153, 209, 215, 231; of natural resources, 6, 47

expulsion, 51–52, 89, 165

extraction, 12, 16–17, 102, 151, 178, 181, 224, 242, 258; of data, 52, 112, 121, 179, 213–14, 216–17

extractivism, 22, 57, 99–100, 126, 151, 153, 212, 215, 223, 242

Facebook (now Meta), 11, 24n23, 183, 192, 199, 203n23

FaceTime, 181

facial recognition systems, xv, 117, 196–98, 209, 259; in landlord technologies, 77, 80, 86–87, 89–90

Fair Credit Reporting Act (1970), 83

Fair Housing Act (1968), 82–83

fake news, 3

family, ix, 51, 71, 85, 108, 142, 148, 173, 178, 246; as caregivers, 152; nuclear, 101, 104, 109; privacy, 106, 138

fascism, 224, 227

Fast Company, 67

Federal Trade Commission, 177

Federation of Women's Health Centers, 180

femininity, 20, 40, 66, 68, 103, 165–66, 219n27; of Alexa, 101, 150; of care work, 153; and transductive labor, 211

feminism, ix, 1, 18, 24n32, 36, 109n4, 127, 232, 260; Black, 142, 180, 209; intervention, 45, 121; Latinx, 180; Marxist, 29; methodology of, 4, 6–8, 13, 16, 38; and property, 80, 89–90; protocol, 180; and Science and Technology Studies (STS), 19, 59, 173, 209; security, 141; and self-knowledge, 20, 174–75, 180; security, 143n9; studies, 12–13, 114, 128, 153, 230; xeno-, 174, 181, 183

fertility, 20, 175–80, 182

Fetz, Eberhard, 257

Fields, Desiree, 85

first-mover disadvantage, 69–70

Fisher, Mark, 23n7

Fitbit, 178

fitness, 20, 175, 178, 234

Flo, 176–79, 182

Flock Safety, 78

Floyd, George, 78, 155

Forbes, 42, 98

Ford Motor Company, 177

Foresta-Inclusive (Tingley), 13, 231–*33*

Forlano, Laura, 161, 169–70

framelessness, 112

Freud, Sigmund, 99, 260n9

Fritsch, Kelly, 156

Fu Manchu, 188

Gabrys, Jennifer, 225

game theory, 69–70, 73

gamification, 195

Garner, Eric, 157n32

gating, 137

gender, 13, 32, 162–63, 167, 170, 193, 239, 244, 248, 259; and capitalism, 3, 10, 12, 238; and care work, 14, 152; and desire, 165, 168; and facial recognition systems, 209; and landlord technologies, 82; and maintenance work, 8, 44–46, 153; and pregnancy-tracking apps, 175; and privacy, 102–3; and race, 150; and rape-reporting apps, 18, 65–66, 69, 72–73; and smart home technologies, 101, 119–21, 127, 134–35, 141, 151, 154; and voice technologies, 206–7, 213, 219n27; and whiteness, 12, 51, 68, 103, 107; and work, 8, 15, 17, 20, 149. *See also* femininity; feminism; heteromasculinity; heteronormativity; masculinity; homophobia; queerness; sexism; sexual violence; transgender people

genocide, 180

gentrification, 84–85, 87, 149

geocode, 69

Gernsback, Hugo, 188

Ghosh, Amitav, 237–38, 244

gig workers, 88, 179

Gillon, Charles, 82

Gilmore, Ruth Wilson, 78

Glabau, Danya, 178

Glaser, April, 65–66

glitches, 22, 94, 108–9, 211

global warming. *See* climate crisis

Glow, 175, 179, 182–83

Google, *12*, 24n23, 188, 203n24, 246; Assistant, 98; Deep Dream Generator, *253*, 255–57; Home, 3; Nest, 77–78, 90, 102, 109

Great Britain, 257; British colonialism, 8, 44

Greenland, 122n9

Grewal, Inderpal, 141, 143n9

Guattari, Félix, 230

Gumbs, Alexis Pauline, 231

Hagelberg, Jacob, 20–21

Haghani, Sanaz: *Close Your Eyes*, 22

Hall, Stuart, 79

hallucinations, 259–60

Hamilton, Jennifer A., 19

Hamraie, Aimi, 151, 156

happiness, 104, 163–64, 166, 170, 183, 248

Happiness Machine, The, 163–64, 166–67

Nelson, Jennifer, 180
Nelson, Robert, 87
neoliberalism, 20, 52–55, 148–50, 154, 203n13, 224, 226–27, 239
Nest, 77–78, 90, 101–2, 109
Netflix, 21, 187, 199
Netscape, 84
neutral mediators, 72
Newsweek, *188*
"new" technologies, 8, 12, 42, 83, 121
New York City, 86–87, 89, 177, 180
New Yorker, 8
New York Post, 70
New York Times, 98–101, 128, 165, 259
Nextdoor, 19, 78, 125–27, 135–42, 143n10
Noel, Vernelle A. A.: *Masks, Mirrors, Light and Shadow*, 22, *159–60*, *168–71*
nonbinary people, 182
nonhumans, 13, 47, 56, 108, 225, 240, 244–45; and AI, xvi, 114, 120–21, 226, 237, 257–58
normalization, 1, 10, 103, 106
normativity, 4–5, 77, 113–14, 148–49, 155, 170, 222, 245; of AI, 20, 115, 117, 166–67; and citizen technoscience, 53–54, 59, 61; facial recognition algorithms and, 86–87; hetero-, 86, 101, 175; of home, 14, 16–17, 19–20; and rape-reporting apps, 70, 73; and techno-creep, 23, 176; and vocal biomarkers, 207–8, 210; of whiteness, 154
Northrop Grumman, 58
#NotMe, 67
Nozick, Robert, 163

Obama, Barack, 67
OceanHill-Brownsville Alliance, 81, 87, 90
Ohlone people, 89
Okafor, Chinyere, 142
Olivares, Jessica L., 19
ontological collectivity, 180
ontology, 6, 180, 208, 249
OpenAI, 259
oppression, 70, 162, 170
optimism, 82, 118, 207
optimization, 71, 118–19, 121, 225; self-, 20, 173, 178, 182
Orwell, George, 200
Orwellian, 200
Osucha, Eden, 103
otherness, 1, 53–55, 59, 126, 128, 167

Our Bodies Ourselves movement, 179–80
outsourcing, 20, 112, 119–20, 174, 181; tedium, 153
Ovia, 178–79

Paasonen, Susanna, 36–37
panaudicon, 21, 207
pandemic. *See* COVID-19 pandemic
Pandemic Shock Doctrine, 10
paramilitary organizations, 18, 54–56. *See also* American Border Patrol (ABP)
paranoia, 126, 139–41, 190
participation, 18, 138, 141, 180, 183, 198, 235; in citizen science, 51–55; in maintenance play, 32, 36; in performance art, 19, 112–17, 119–20; in rape-reporting apps, 70–71; in transductive research, 215, 217
Parvin, Nassim, 51, 174
Pasquinelli, Matteo, 117
patents, 50, 58, 126–27, 142; for home security, 19, 128–36
paternalism, 18, 80, 99
pathological speech, 210
patriarchy, 70, 101, 106–7, 151, 175
PayPal, 175
PBS News Hour, 67
Pence, Mike, 201n3
[people.power.media], 81
People's Free Medical Clinics, 180
period tracking. *See* menstruation-tracking apps
Phan, Thao, 151, 153, 219n23
phatic labor, 211
piggy banks, 103–6
Pichai, Sundar, 203n24
Pipecleaner (Behar), *45–46*
Pister, Kristofer, 43–44, 46
plants, 138, 225, 241–42, 244–45; in COVID times, 238–39, 246–49, 256; creep of, 4, 7, 21, 226, 235, 240, 243, 249–50, 255
platform(s), 1, 18–19, 45, 59, 73, 83, 99, 176, 182–83, 259–60; American and Chinese tech, 193, 195–96, 198–200; and creep, 5, 9–10, 16, 174; for Internet of Things, 224; labor, 28; in landlord technologies, 79–80, 85, 87; Nextdoor, 126, 135, 137, 141; privacy of, 12; for rape reporting, 69; schooling, 78; social media, xv, 183, 190; webcam, 17, 28–29, 31–35, 37–38

play, 21–22, 40, 54, 84, 94, 119, 149, 168–69, 217, 256–58; with categories, 239–40, 249–50; maintenance and, 17, 28, 35–38; music, 150–53; senses at, 46; work and, 246, 249

police: and Black Lives Matter movement, 88, 155–56; police violence, 143n16, 157n32; and Nextdoor vigilantism, 125, 128, 134, 138–42; and sexual violence cases, 66–68, 70, 72; and smart dust, 43

policing, 79, 106, 134, 217, 230; and citizen scientists, 53, 55, 59; of landlord technologies, 9, 80–81; racialized, 10, 12, 18, 107, 154

pollution, 46, 225, 247

porch pirates, 138

Porkfolio, 103–5

pornography, 31–36

possession, xvi, 240; dis-, 18, 79–80, 82, 84–85, 88, 103, 217; self-, 3, 106–7

Postman, Neil, 203n26

post-9/11 era, 79, 84

poverty, 83, 87, 137, 152, 213

Povinelli, Elizabeth, 242

Power, Emma, 82

pregnancy-tracking apps, 16, 20, 173–84

prison(s), 67, 98, 217; digital, 80; industrial complex, 78

privacy, xv, 24n23, 144n34, 151, 208, 224, 246, 248; and Cherry Home, 14–15, 97; as commodity, 19, 107; in COVID times, 239; loss of, 1, 3, 98, 191; and Nextdoor, 127, 137; and pregnancy-tracking apps, 174–75, 177, 180; as property, 12, 103, 144n34; right to, 3, 5, 9, 10–12, 102–3, 106–7, 113; and smart dust, 42; and smart home technologies, 102, 149–50, 155; and surveillance binary, 3, 107, 113, 174, 207

privatization, 81, 83, 88, 150, 156, 240

privilege, 38, 105, 187, 192, 197, 255; of citizen, 51, 59; during COVID-19 pandemic, 10, 99; in gated communities, 137; of individualism, 5, 9–10; in landlord technologies, 18, 80; of privacy, 10, 12, 103, 113; reproductive, 107, 175; of sight, 5, 23, 47; of whiteness, 10, 17, 71, 103, 106

productivity, 14, 20, 114, 152, 165, 167–68, 173–75, 177–79; non-, 33, 162, 225

property, 6, 101, 108–9, 222; and landlord tech, 80–82, 85–88; ownership, 9–10, 107, 126–27;

protection of, 11, 78–79, 130, 134–35, 137, 140–42; racialized, 5, 18, 78–79, 103–4, 107, 127, 144n34; and surveillance vigilantes, 19, 138, 140–41; undoing, 80, 88–89; and whiteness, 12, 17, 88, 127–28, 141

prophetic vision, 21, 188, 194

Proposition 187, 50

proptech, 80–81, 84, 86. *See also* landlord technologies

protocol feminism, 180

psychiatry, 206, 208, 218

psychotropic houses, 94, 109

public health, 177–78, 181, 207, 246

public science, 54

Qoobo, 1–2, 5

quantification, 1, 10, 17, 79, 108, 178, 212

queerness, 38, 67–68, 90, 142, 182, 245

Quirky, 103–6

racial banishment, 89

racial capitalism, 10, 17, 101, 127, 150–51, 179, 248–49, 260; in border security, 60; technologically augmented, 4, 6, 78, 83

racialism, 79

racial profiling, 126

racial regimes of ownership, 79

racial technocapitalism, 79–81, 84, 87, 89

racism, 126, 144n34, 157n32; in AI, 121, 242, 259; anti-, 18, 83, 90, 180; anti-Black, 128; anti-immigration, 50–51, 59; in housing industry, 78, 82–84, 87, 90; medical, 181–82; structural, 231, 259–60; white supremacy, 66, 78–79, 142, 151, 209

radical care, 181, 183–84, 214

Rainer, Yvonne: *The Mind Is a Muscle*, 33

Ralph, Michael, 179

Rand Corporation, 57

rape-reporting apps, 16, 18, 67, 69, 71–72

rationality, 11, 20, 54, 154, 161–62, 165–66, 209, 233, 245; of AI, 255, 257

Ray, Kasturi, 179

real estate industry, 80, 84–85, 222. *See also* landlordism

realism, xvi, 166

Rebeiz, Linda Dounia, 259

reclamation, 89

redlining, 81, 126

reductionism, 47, 102, 107, 117–18, 170, 177–78, 209–11, 259–60; of humans, 120–21 of property, 141–42; and technological creep, 3, 5, 7, 9, 13, 126

rematriation, 88–89

reproduction, 20, 55, 60, 79, 103–4, 107, 173, 175; of maintenance work, 35, 37, 102; social, 19, 101, 153, 174

reproductive labor, 29, 33–35, 152–53, 179, 245, 248; apps for, 174–75, 178; rights of, 107, 180

resilience, 8, 40, 46

resistance, 3, 5, 7–8, 16, 21–23, 142, 148, 156, 211

Resisters, The (Jen), 97–98, 107–8

responsibility, 42, 50, 61, 174; of AI, 116, 118, 242; of citizenship, 53–55, 59, 105; of consumers, 10, 178; in research, 214, 217–18; of women, 119, 234–35

Ring, 100, 101, 126, 128, 130, 135–36, 143n16

Robinson, Cedric, 79

Robinson, Dylan, 217

robots, xvi, 49, 255; as creepy, 9, 14, 166; and Qoobo, 1–2, 5; and robotics, 17, 164; in transductive labor, 213

robust living, 234

Roe v. Wade, 174, 177

Rogan, Joe, 201n3

Rogers, Fabian, 87

Rohmer, Sax, 188

Rojas, David, 224–25, 227

Roose, Kevin, 165–68

Roy, Ananya, 80, 89

Roy, Arundhati, 248

Roy, Deboleena, 230

Sanders, Ron, 60

Schultz-Figueroa, Benjamin, 94

Science and Technology Studies (STS), 155; feminist, 19, 59, 173, 209

science fiction, 83, 94, 188, 190, 200

security, 239; border, 54–57, 60–61; data, 107, 177; at home, 19, 98, 100, 126–31, 134–38, 140–42; and landlord technologies, 78, 86–87

seismic detection and ranging mechanism (SEIDARM), 58, 61

self, 1, 40, 105, 142, 163, 166

self-care, 20, 70, 173–74, 177–81

self-driving cars, 3, 10, 231

self-knowledge, 174–75, 178, 180–81, 184

self-optimization, 20, 178, 182

self-policing, 106

self-possession, 3, 105–7

self-reliance, 54–55

self-tracking, 9, 20, 173–84

Semel, Beth, 21

semiconductor industry, 224

Sen, Mithu, 21, 240, 246, 248–49; *UnLOCKDOWN*, 241; *UnMYthU: UnKIND(s) Alternatives* (Sen), 241–44

sensationalization, 3, 8, 57, 98, 162, 245

sensors, xi, 21, 112, 206, 239; in border enforcement, 49–50, 56, 58–59; networked tooth, 1–3; in smart dust, 8, 41, 43–44, 46–47; of smart forests, 7, 13, 224–27, 231–32; in smart homes, 14, 97, 108, 112, 148

Sesame Credit, 195–*96*, 198, 202n13, 202n16

settler colonialism, 6, 17–18, 51, 55, 66, 217, 224, 234, 259; fantasies of, 58; and property, 88, 107; and white supremacy, 142

sexism, 65, 119, 121, 150, 242

sexual violence, 16, 18, 65–72, 219n23

sex webcam (sexcam) platforms, 17, 28–29, 31–38

sex work, 17, 28–29, 45–46, 68

Sharma, Sarah, 121

Shaw, Julia, 71–72

Shelby, Renee, 18

Shell, 47

signal processing, 206

Silicon Valley, 83, 150, 175, 196, 198–200, 260

Sinophobia, 188, 199–200

SkyBell, 130, 135, 143n16

slavery, 6, 12, 88, 102, 107, 120, 127, 144n34, 179, 231

small unmanned aircraft system (sUAS), 50, 55, 58

smart cities, 222, 224, 230–31

smart desires, 20, 162, 165, 170

smart dust, 8, 17, 41–48, 59

smart forests, 7, 13, 17, 21, 223–35

smart home assistants, 17, 121, 148, 154, 241; femininity of, 119, 150; and *LAUREN* (McCarthy), 19, 111–17, 119–20. *See also* Alexa

smart homes, 9, 22, 94–95, 107, 109, 114; and Alexa, 148; as creepy, 98–100, 148–49; and hometech, 101, 104, 126, 130; and people with disabilities, 19–20, 154; prevalence of, 85, 98; and privacy, 12, 19, 98, 102–3 smart wives, 119

snitching apps, 78

social credit imaginary, 190–91, 196–97

Social Dilemma, The, 199

social media, xv, 3, 9, 162, 168, 197, 199; as creepy, 190; and hometech, 135–36; and Mithu Sen, 240–41; Nextdoor, 19, 125–26, 137, 141; and rape-reporting apps, 71; use in COVID times, 247–48. See also *individual platforms*

social reproduction, 19, 101, 153, 174

Sogorea Te' Land Trust, 89

solidarity, 18, 71, 90, 156, 176, 239

Soros, George, 201n3

Soscia, Ryan, 70

sound studies, 210

sousveillance, 13, 24nn30–31, 89

Southern Poverty Law Center, 50

sovereignty, 52–53, 55, 59–60

spectrogram, 210

Spencer, Glenn, 50, 55–56, 58–60

Spot AI, 18

Spotify, 153

spying, 7, 100, 126

Squire, Megan, 245

Srnicek, Nick, 14

Stability AI, 259

Stallabrass, Julian, 258

Star, Susan Leigh, 35

Stauffer, Robert, 83

stereotypes, 68, 119, 150, 153, 175, 240

sterilization, 107, 180

StoneLock, 87–88, 90

Street Smarts (Bennett), 22, *221–22*, *230–31*

Stewart, Whitney Nell, 144n34

Strengers, Yolande, 119

subprime mortgage crisis (2008), 80, 84

subversion, 21, 30, 34, 37, 108, 207, 212, 240–41, 248–49

Suchman, Lucy, 258

SuperCalc, 83

surrealism art movement, 256–58

surrogate human effect, 53

surveillance, 1, 148, 154, 222, 240, 246; and ad-tech, 181; and border technologies, 17; capitalism, 10, 24n23, 102, 150, 178; in China, 196–98; and creep, 3, 5, 7–11, 16, 249; culture, 148; digital epidemiological, 246; in domestic space, 13–15, 18, 101, 108–9, 113, 135, 154; and health apps, 174–75, 182; and landlord tech, 78–82, 86–88; and maintenance play, 36; neighborhood, 78, 126, 137–40; and panaudicon, 207; patent, 129, 132–34; and policing, 18; in pregnancy-tracking apps, 20; and privacy binary, 16, 107, 113, 174, 207; racialized, 78–79, 106, 113, 182; self-care as, 179; of sound, 209; studies, 12–13, 128; vigilantes, 19, 125, 127, 137–38, 141–42; and vocal biomarker research, 216, 218

survival, 149, 154, 183, 217, 249; in Black feminist thought, 142; creep as, 21–22; and empowerment, 73; and sexual violence, 18, 65–73

sustainability, 52, 100, 223

Sutter, John D., 41

symbolic annihilation, 183

Takeshita, Chikako, 78

Talk to Spot, 67, 69, 71–73

Taylor, Breonna, 143n16

techlash, 190

technoaffective attunement, 217

technocapitalism, 3, 5, 8, 21–22, 162, 260; racial, 79–81, 84, 87, 89

technocultures, 191, 194, 244

techno-empowerment, 69

techno-evangelism, 148

technofixes, 20, 149, 217

technoliberalism, 61, 119–21, 150, 152–53, 155–56, 199, 248

technological determinism, 112, 118, 121, 210

technonationalism, 20, 190, 199–200

techno-optimism, 118, 207

techno-Orientalism, 119–21, 188, 190, 200, 201n3

technophilia, 209

technophobia, 209

technoscience, 18, 118, 207, 209–11, 217, 257; citizen, 52, 54–56, 58, 60; creep of, 21, 156; feminist, 180; inequalities of, 150, 154; uncivil, 51–52, 54–61

technosolutionism, 67, 121

technostress, 246

techno-utopianism, 21, 191, 223

tenants: activism by, 9, 18, 80–81, 85–87, 89–90; screening of, 78, 80–86, 88–89

territory, 17, 36, 43–44, 50–52, 55, 59, 222

Walcott, Rinaldo, 88
war, 17, 84, 128, 247; attachments to, 154;
    Cold War, 54, 79, 199
Warren, Samuel, 102–3
Washington Post, 182, 193–94
waste, 8, 46, 215, 224
Waze, 109
wearable technology, xv, 20, 175
webcams, 17, 28, 32. *See also* sex webcam
    (sexcam) platforms
Weinbaum, Alys Eve, 38n2
Weinstein, Harvey, 65
weird, the, 23n7
Weizenbaum, Joseph: ELIZA, 123n36
WeWork, 84
What to Expect (app), 176
White House Office of Science and
    Technology, 67
whiteness, 219n27; creep of, 18, 68–69, 72; and
    gated communities, 137; policing uphold-
    ing, 18, 141; privileges of, 17, 71, 103, 106–7;
    and property, 12, 17, 88, 127–28, 141; and
    rape-reporting apps, 67–71, 73; of smart
    home technologies, 101, 103, 151, 154
white supremacy, 66, 78–79, 142, 151, 209
white-washing, 150
Whyte, Kyle Powys, 234
Wiehn, Tanja, 19
Wiener, Norbert, 210
Winger-Bearskin, Amelia, *117*
Wired, 65, 128
Wizard of Oz (WoZ) setup, 212–14, 217
Wolomi (app), 182
Women in Proptech, 80
women's rights, 65–66. *See also* feminism;
    #MeToo movement

Woods, Heather Suzanne, 150
work, 47, 87, 89, 102–4, 128, 142, 176, 180; and
    Asian trope, 120; and *Black Mirror* episode,
    191–93, 198; care, 14, 59, 151–53, 174, 206–7;
    of citizen scientists, 51–52, 54–56, 58, 61; in
    COVID-19 times, 245–47; and environ-
    mentalism, 225, 227, 230; and exploitation,
    8; of feminists, 12, 114; gendered, 15, 153;
    and harassment, 69–73; of home security
    systems, 130, 132; maintenance, 8, 17, 30,
    32–36, 44–46, 106; productivity, 14, 174,
    177–79; reproductive, 33, 35, 248; and self-
    tracking apps, 183; sex, 17, 28–29, 35, 45–46;
    and smart home technologies, 154, 156, 242;
    and technocapitalism, 5–6, 79–82, 89; with
    technocreep, 238–40; of transductive labor,
    21, 208–17. *See also* artwork
Wosk, Julie, 164

X. *See* Twitter/X
xenofeminism, 174, 181, 183–84
xenophobia, 16, 50–51, 242

Yale University, 182
Yamin, Priscilla, 106
Yang, Wayne, 105
Yardi, 85
Yelp, 192
Yolsu Indigenous Homeland, 233
Young Lords, 180

Zimmerman, George, 137
Zoom, ix, 181, 239, 246–47
Zuboff, Shoshana, 24n23, 102, 108, 178
Zuckerberg, Mark, 199
Zuijderduijn, Jaco, 106